Management of Data Center Networks

Management of Data Center Networks

Edited by

Nadjib Aitsaadi
Universités Paris-Saclay, UVSQ, DAVID, F-78035, Versailles, France

IEEE Press
Series on
Networks and
Services Management

Dr. Veli Sahin and
Dr. Mehmet Ulema, *Series Editors*

IEEE PRESS

WILEY

Published by John Wiley & Sons, Inc., Hoboken, New Jersey.
Published simultaneously in Canada.

For general information on our other products and services or for technical support, please contact our Customer Care Department within the United States at (800) 762-2974, outside the United States at (317) 572-3993 or fax (317) 572-4002.

Wiley also publishes its books in a variety of electronic formats. Some content that appears in print may not be available in electronic formats. For more information about Wiley products, visit our web site at www.wiley.com.

Library of Congress Cataloging-in-Publication Data applied for:

ISBN: 9781119647423

Cover Design: Wiley
Cover Image: © Bill Donnelley/WT Design

Set in 9.5/12.5pt STIXTwoText by SPi Global, Chennai, India

To my daughters L. H. S.

Contents

About the Editor

Nadjib Aitsaadi, PhD, is a Full Professor in Networks and Telecommunications at UVSQ Paris-Saclay University in France. He is a member of DAVID Laboratory and leads the Next Generation Networks Team. Prof. Aitsaadi earned a PhD in 2010 from Sorbonne University in Networks & Telecommunications and graduated in 2016 with a "Habilitation" diploma from University Paris Est (UPE). His main research fields are the security and QoS optimization of cellular networks (5G, 6G, HAP), IoT, DCN, V2X, MEC, NFV/SDN, and more. The results have been published in many major journals such as IEEE JSAC, IEEE TVT, Elsevier ComNet, Elsevier ComCom, etc. and major conferences such IEEE SECON, IEEE LCN, IEEE MASS, ACM MSWiM, IEEE/IFIP NOMS/IM, IEEE ICC, IEEE GLOBE-COM, etc. Prof. Aitsaadi chairs many tracks in IEEE/IFIP conferences such as IEEE GLOBECOM, IEEE/IFIP IM, IEEE/IFIP CIoT, etc. and he is very active in IEEE Technical Committees TCIIN.

Contributors

Nadjib Aitsaadi
Universités Paris-Saclay, UVSQ
DAVID
Versailles
France

Dallal Belabed
Airbus Defense and Space, Airbus
Saint-Quentin en Yvelines
Elancourt
France

Selma Boumerdassi
CEDRIC, CNAM
Paris
France

Boutheina Dab
VMware, Hauts-de-Seine
La Defense
France

and

LiSSi Lab, UPEC, Val de Marne
Vitry sur Seine
France

Ilhem Fajjari
Orange Labs, Orange
Hauts-de-Seine
Chatillon
France

Ruben Milocco
GCAyS, UNComahue
Neuquen
Argentina

Pascale Minet
Inria
Paris
France

Eric Renault
LIGM, Univ. Gustave Eiffel, CNRS
ESIEE Paris
Marne-la-Vallée
France

Oussama Soualah
OS-Consulting
Athis-Mons, Essonne
France

and

LiSSi Lab, UPEC, Vitry sur Seine
Val de Marne
France

Acronyms

DCN	Data Center Network
ECMP	Equal Cost Multi-Path
GMPLS	Generalized Multi-Protocol Label Switching
HDCN	Hybrid Data Center Network
IGP	Interior Gateway Protocol
LISP	Locator/Identifier Separation Protocol
MSDC	Massively Data Center
MPTCP	Multipath TCP
NFV	Network Function Virtualization
OSPF	Open Shortest Path First
QoS	Quality of Service
SCTP	Stream Control Transmission Protocol
SDN	Software Defined Network
SN	Substrate Network
STP	Spanning Tree Protocol
STT	Stateless Transport Tunneling
TCI	Tag Control Information
TE	Traffic Engineering
ToR	Top of Rack
TPID	Tag Protocol IDentifier
TRILL	Transparent Interconnection of Lots of Links
VDC	Virtual Data Center
VLC	Visible Light Communication
VNE	Virtual Network Embedding
VXLAN	Virtual Extensible LAN
WTU	Wireless Transmission Unit

Introduction

Thanks to the advent of the long-awaited fifth-generation (5G) mobile networks, mobile data and online services are becoming widely accessible. Discussions of this new standard have taken place in both industry and academia to design this emerging architecture. The main near future objective is to ensure the capability to respond to the different applications needs such as videos, games, web searching, etc., while ensuring a higher data rate and an enhanced Quality of Service (QoS). While no official standardization is yet delivered for 5G, experts assure that, the impressive proliferation of smart devices will lead to the explosion of traffic demand. Billions of connected users are expected to deploy a myriad of applications.

In this respect, recent statistics elaborated by CISCO Visual Networking Index (VNI) highlight that the annual global IP traffic will roughly triple over the next 5 years and will reach 2.3 zettabytes by 2020. More specifically, it is expected that smart phones traffic will impressively increase from 8% in 2015 to 30% of the total of IP traffic in 2020. Mobile data traffic per month will grow from 7 Exabytes in 2016 to 49 Exabytes by 2021. In particular, tremendous video traffic will be crossing IP networks to reach 82% of the totality of IP traffic. It is also expected that the number of connected mobile devices will be more than three times the size of the global population by 2020. In this regard, future networks are anticipated to support and connect plenty of devices, while offering higher data rate and lower latency.

To cope with this unprecedented traffic explosion, the service providers are urged to rethink their network architectures. In fact, efficient scalable physical infrastructures, e.g., Data Centers (DCs), are required to support the drastically increasing number of both online services and users.

To manage their DCs infrastructure, many of giant service tenants are resorting to virtualization technologies making use of Software Defined Networking (SDN) and Network Functions Virtualization (NFV).

On the one hand, SDN controllers offer the opportunity to implement more powerful algorithms thanks to a real-time centralized control leveraging an accurate view of the network. Indeed, thanks to the separation of the forwarding and the control planes, the managements' complexity of the network infrastructure is considerably reduced while providing tremendous computational power compared to legacy devices. On the other hand, thanks to NFV paradigm, network functions and communication services are first softwarized and then cloudified, so that they can be on demand orchestrated and managed as cloud-native IT applications. It is straightforward to see that these approaches are complimentary. They offer a new way to design and manage data centers while guaranteeing a high level of flexibility and scalability. The new emerging SDN and NFV technologies requires scalable infrastructures. To that end, a great deal of efforts have been devoted to the design of efficient DC architectures. Indeed, Internet giants ramped up their investment in data centers/IT infrastructures and poured in billions of dollars to widen their global presence and improve their competitiveness in the Cloud market.

In this context, the latest Capital Expenditure (CAPEX) of the five largest-scale Internet operators, Apple, Google, Microsoft, Amazon, and Facebook, increased by 9.7% in 2016 in order to invest in designing their DCs. Over the past years, these companies have spent, in total, a capital of 115 $ billions, to build out their DCs. For instance, Google has invested millions of dollars in expanding its data centers spread all over the world: Taiwan, Latin America, Singapore, etc. Facebook has started, since 2010, building out its own DCs in Altoona, Iowa, and North Carolina. In this regard, efficiently designing data centers is a crucial task to ensure scalability required to meet today's massive workload of Cloud applications. Moreover, it is mandatory to deploy the proper mechanisms for routing and resource allocation to communication flows in DCs.

To deal with these challenges, we investigate, in this book, a radically new methodology changing the design of traditional Data Center Network (DCN) while ensuring scalability and enhancing performance. First, in Chapter 1, we will overview the Data Center network architecture. Then, in Chapter 2, we will summarize the main related DC networks routing strategies at layer-2, layer-3, and up layers. Besides, an overview of Traffic Engineering (TE) techniques from link-state while considering TCP fairness models. Next, in Chapter 3, we will overview the related work addressing intra-data center resource allocation and routing in both wired and/or wireless (i.e., hybrid) data center networks. Afterwards, in Chapter 4, we will summarize the main reliable virtual network embedding strategies connecting geographically distributed data centers. Thanks to network function virtualization, CAPEX and OPEX are deeply reduced while ensuring the requested quality of service is more complex. Finally, in Chapter 5, we will provide a methodology to evaluate the energy cost reduction in DC brought by proactive management, while keeping a high level of user satisfaction.

1

Architectures of Data Center Networks: Overview

Boutheina Dab[1,2], Ilhem Fajjari[3], Dallal Belabed[4], and Nadjib Aitsaadi[5]

[1] *VMware, Hauts-de-Seine, La Defense, France*
[2] *LiSSi Lab, UPEC, Val de Marne, Vitry sur Seine, France*
[3] *Orange Labs, Orange, Hauts-de-Seine, Chatillon, France*
[4] *Airbus Defense and Space, Airbus, Saint-Quentin en Yvelines, Elancourt, France*
[5] *Universités Paris-Saclay, UVSQ, DAVID, F-78035, Versailles, France*

Abstract

To deal with the widespread use of cloud services and the unprecedented traffic growth, the scale of the Data Center has importantly increased. Therefore, it is crucial to design novel efficient network architectures able to satisfy the requirements on bandwidth. As a key physical infrastructure, Data Center Network (DCN) designing has widely been a hot research focus.

This chapter reviews the main DCN architectures propounded in the literature. To do so, a taxonomy of DCN designs will be proposed, while analyzing in depth each structure of the given classification. Then, we will provide a qualitative comparison between these different DCN groups. Finally, we will present hybrid DCN architecture based on wired and wireless architecture.

1.1 Taxonomy of DCN Architectures

In this section, we present a taxonomy of the existent Data Center Network (DCN) architectures with a detailed review of each drawn class. In general, several criteria have to be considered to design robust DCNs, namely, high network performance, efficient resource utilization, full available bandwidth, high scalability, easy cabling, etc. To deal with the aforementioned challenges, a panoply of solutions have been designed. Mainly, we can distinguish two research directions. In the first one, wired DCN architectures have been upgraded to build advanced cost-effective topologies able to scale up data centers. The second approach has resorted to deploying new network techniques within the existing DCN so as to handle the challenges encountered in the prior architectures. Hereafter, we will give a detailed taxonomy of these techniques.

1.1.1 Classification of DCN Architectures

With regard to the aforementioned research directions, we can identify three main groups of DCN architectures, namely, **switch-centric** DCN, **server-centric** DCN, and **enhanced** DCN. Each group includes a variety of categories that we will detail hereafter.

- *Switch-centric DCN architecture*: switches are, mostly, responsible for network-related functions, whereas the servers handle processing tasks. The focus of such a design is to improve the topology so as to increase network scale, reduce oversubscription, and speed up flow transmission. Switch-centric architectures can be classified into three main categories according to their structural properties:
  1. *Traditional tree-based DCN architecture*: represents a specific kind of switch-centric architecture, where switches are linked in a multirooted form.
  2. *Hierarchic DCN architecture*: is a switch-centric DCN, where network components are arranged in multiple layers. Each layer characterizes traffic differently.
  3. *Flat DCN architecture*: compresses the three switch layers into only one or two switch layers, in order to simplify the management and maintenance of the DCN.
- *Server-centric DCN architecture*: servers are enhanced to handle networking functions, whereas switches are used only to forward packets. Basically, servers are simultaneously end-hosts and relaying nodes for multihop communications. Usually, server-centric DCN are recursively defined multilevel topologies.
- *Enhanced DCN architecture*: is a specific DCN which is tailored for future Cloud computing services. Indeed, the future research direction attempts to deploy networking techniques so as to deal with wired DCN designs limitations. Recently, a variety of technologies have been used in this context, namely, optical switching and wireless communications. Accordingly, we distinguish two main classes of enhanced DCN architectures:

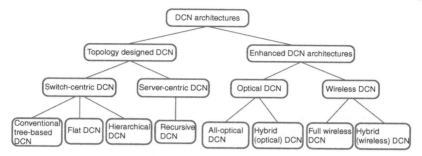

Figure 1.1 Taxonomy of DCN architectures.

1. *Optical DCN*: makes use of optical devices to speed up communications. It can be either: (i) all-optical DCN (i.e. with completely optical devices) or (ii) hybrid optical DCN (i.e. both optical and Ethernet switches).
2. *Wireless DCN*: deploys wireless infrastructure in order to enhance network performance, and may be: (i) fully wireless DCN (i.e. only wireless devices) or (ii) Hybrid DCN (i.e. both wireless and wired devices).

Figure 1.1 illustrates the taxonomy of current DCN architectures. In the following, we will detail each category and discuss their impact on Cloud computing performance.

1.1.2   Switch-Centric DCN Architectures Overview

1.1.2.1   Tree-Based DCN

The traditional DCN is typically based on a multiroot tree architecture. The latter is a three-tier topology composed by three layers of switches. The top level (i.e. root) represents the core layer, the middle level is the aggregation layer, while the bottom level is known as the access layer. The core devices are characterized by high capacities compared with aggregation and access switches. Typically, the core switches' uplinks connect the data center to the Internet. On the other hand, the access layer switches commonly use 1 Gbps downlink interfaces and 10 Gbps uplink interfaces, while aggregation switches provide 10 Gbps links. Access switches (i.e. top of rack, ToRs) interconnect servers in the same rack. Aggregation layer allows the connection between access switches and the data forwarding. It is worth noting that the above values of network interface cards throughput are continuously increasing. For instance, nowadays it is easy and not really expensive

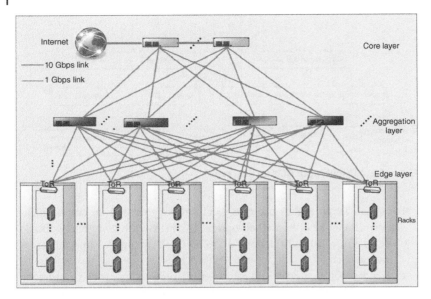

Figure 1.2 Traditional tree-based DCN architecture.

to deploy interfaces with 25 and 100 Gbps. An illustration of tree-based DCN architecture is depicted in Figure 1.2.

Unfortunately, traditional DCNs struggle to resist to the increasing traffic demand. First, core switches are prone to bottlenecks issues as soon as the workloads reach the peak. Moreover, in such a DCN, several downlinks of a ToR switch share the same uplink which limits the available bandwidth. Second, DCN scalability strongly depends on the number of switch ports. Therefore, the unique way to scale this topology is to increase the number of network devices. However, these solutions results in high construction costs and energy consumption. Third, tree-based DCN suffers from serious resiliency problems. For instance, if a failure happens on some of the aggregation switches, then servers are likely to lose connection with others. In addition, resource utilization is not efficiently balanced. For all the aforementioned reasons, researchers put forward alternative DCN topologies.

1.1.2.2 Hierarchical DCN Architecture

Hierarchical topology arranges the DCN components in multiple layers. The key insight behind this model is to reduce the congestion by minimizing the oversubscription in lower-layer switches using the upper-layer devices. In the literature, we find several hierarchic DCN examples, namely, CLOS, FatTree, and VL2. Hereafter, we will describe each one of them.

CLOS-Based DCN Is an advanced tree-based network architecture. It was, first, introduced by Charles Clos, from Bell Labs, in 1953 to create nonblocking multistage topologies, able to provide higher bandwidth than a single switch. Typically, CLOS-based DCNs come with three layers of switches: (i) access layer (ingress), composed of the ToRs switches, directly connected to servers in the rack; (ii) aggregation layer (middle), formed by aggregation switches referred as spines and connected to the ToRs; and (iii) core layer (egress), formed by core switches serving as edges to manage traffic in and out the DCN (Chen et al., 2016).

The CLOS network has been widely used to build modern IP fabrics, generally referred to as spine and leaf topologies. Accordingly, in this kind of DCN, commonly named folded-CLOS topology, the spine layer represents the aggregation switches (i.e. spines), while the leaf layer is composed of the ToR switches (i.e. leaves). In other words, in CLOS topology, (i) leaf layer is composed of ToR switches and (ii) spine layer is composed of aggregation switches. The spine layer is responsible for interconnecting leafs. CLOS inhibits the transition of traffic through horizontal links (i.e. inside the same layer). Moreover, CLOS topology scales up the number of ports and makes possible huge connection using only a small number of switches. Indeed, augmenting the switches ports enhances the spine layer width and, hence, alleviates the network congestion. In general, each leaf switch is connected to all spines. In other words, the number of up (respectively, down) ports of each ToR is equal to the number of spines (respectively, leaves). Accordingly, in a DCN of n leaves and m spines, there are $n \times m$ wired links. The main reason behind this link redundancy is to enable multipath routing and to mitigate oversubscription caused by the conventional link state open shortest path first (OSPF) routing protocol. In doing so, CLOS network provides multiple paths for the communication to be switched without being blocked.

CLOS architecture succeeds to ensure better scalability and path diversity than conventional tree-based Data Center (DC) topologies. Moreover, this design reduces bandwidth limitation in aggregation layer. However, this architecture requires homogeneous switches and deploys huge number of links.

Fat-Tree DCN Is a special instance of CLOS-based DCN introduced by Al-Fares et al. (2008) in order to remedy the network bottleneck problem existing in the prior tree-based architectures. Specifically, Fat-Tree comes with a new way to interconnect commodity Ethernet switches. Typically, it is organized in k pods, where each pod contains two layers of $k/2$ switches. Each k-port switch in the lower layer is directly connected to $k/2$ hosts, and to $k/2$ of the k ports in the aggregation layer. Therefore, there is a total of

$(k/2)^2$ k-port core switches, each one is connected to each port of the k pods. Accordingly, a Fat-Tree built with k-port switches supports $k^3/4$ hosts.

The main advantage of the Fat-Tree topology is its capability to deploy identical cheap switches, which alleviates the cost of designing DCN. Further, it guarantees equal number of links in different layers which inhibits communication blockage among servers. In addition, this design can importantly mitigate congestion effects thanks to the large number of redundant paths available between any two given communicating ToR switches. Nevertheless, Fat-Tree DCN suffers from complex connections, and its scalability is closely dependent on the number of switch ports. Moreover, this structure is impacted by the possible lower-layer devices failure which may entail the degradation of DCN performance.

This architecture has been improved by designing new structures based on a Fat-Tree model, namely, **ElasticTree** Heller et al. (2010), **PortLand** Mysore et al. (2009), and **Diamond** Sun et al. (2014). The main advantage of such topologies is to reduce maintenance cost and enhance scalability by reducing the number of switch layers.

Valiant Load Balancing DCN Architecture Valiant load balancing (VLB) is introduced in order to handle traffic variation and alleviate hotspots when random traffic transits through multipaths. In the literature, we find, mainly, two kinds of VLB architectures. First, **VL2** is three-layer CLOS architecture introduced by Microsoft in Greenberg et al. (2009a). Contrarily to Fat-Tree, VL2 resorts to connecting all servers through a virtual two-layer Ethernet, located in the same local area network (LAN) with servers. Moreover, VL2 implements VLB mechanism and OpenFlow to perform routing while enhancing load balancing. To forward data over multiple equal cost paths, it makes use of equal-cost multi-path (ECMP) protocol. VL2 architecture is characterized by its simple connection and does not require software or hardware modifications. Nevertheless, it still suffers from scalability issue and does not take into account reliability, since single node failure problem persists.

Second, **Monsoon** architecture (Greenberg et al., 2008), aims to alleviate over-subscription based on a two-layer network that connects servers and a third layer for core switches/routers. Unfortunately, it is not compatible with the existing wired DCN architecture.

1.1.2.3 Flat DCN Architecture

The main idea of the Flat switch-centric architectures is to flatten down the multiple switch layers to only two or one single layer, so as to simplify maintenance and resource management tasks. There are several topologies that are proposed for this kind of architecture. First, the authors of Abts et al. (2010) conceive flattened butterfly (**FBFLY**) architecture to build energy-aware DCN. Specifically, it considers power consumption proportionally to the traffic load, and so replaces the 40 Gbps links by several links with fewer capacity regarding the requested traffic in each scenario. colored butterfly (**C-FBFLY**) Csernai et al. (2015) is an improved version of FBFLY which makes use of the optical infrastructure in order to reduce cabling complexity while keeping the same control plane. Then, **FlaNet** Lin et al. (2012) is also a two-layer DCN architecture. Layer 1 includes a single n-port switch connecting n servers, whereas the second layer is recursively formed by n^2 one-layer FlatNet. In doing so, this architecture reduces the number of deployed links and switches by roughly $1/3$ compared to the classical three-layer FatTree topology, while keeping the same performance level. Moreover, FlatNet guarantees fault-tolerance thanks to the two-layer structure and ensures load balancing using the efficient routing protocols.

Discussion In conclusion, switch-centric architectures succeed to relatively enhance traffic load balancing. Most of these structures ensure multirouting. Nevertheless, such a design brings up in general at least three layers of switches which strongly increases cabling complexity and limits, hence, network scalability. Moreover, the commodity switches commonly deployed in these architectures do not provide fault-tolerance compared to the high-level switches.

1.1.3 Server-Centric DCN Architectures Overview

In general, these DCN architectures are conceived in a recursive way where a high-level structure is formed by several low-level structures connected in a specific manner. The key insight behind this design is to avoid the bottleneck of a single element failure and enhance network capacity.

The main server-centric DCN architectures found in the literature include **BCube**, which is a recursive server-centric architecture (Guo et al., 2009a) that makes use of on specific topological properties to ensure custom routing protocols. Another one is **DCell** which is a recursive architecture built on switches and servers with multiple network interface cards (NICs) (Guo et al., 2008b). The objective is to increase the scale of servers. Finally, **CamCube** Abu-Libdeh et al. (2010) is a free-of-switching DCN architecture, specifically modeled as a 3D DCN topology, where each server connects to exactly two servers in 3D directions, however, the suppression of the swathing by a direct connection between servers seems unfeasible.

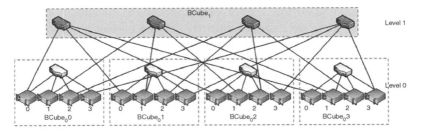

Figure 1.3 BCube$_1$ with $n = 4$.

In the following, we will cover the DCell and the BCube architectures since they represent the most cited ones.

The BCube Guo et al. (2009b) topology is a recursive architecture designed for shipping and container-based, modular data center (MDC). As depicted in Figure 1.3, the BCube solution has server devices with multiple ports (typically no more than four). Multiple layers of cheap commodity off-the-shelf miniswitches are used to connect those servers. A BCube$_0$ is composed of n servers connected to an n-port switch. A BCube$_1$ is constructed from n BCube$_0$s and n n-port switches. More generally, a BCube$_k$ ($k \geq 1$) is constructed from n BCube$_{k-1}$s and n^k n-port switches. For example, in a BCube$_k$ with n n-port switch, there are $k + 1$ levels of switches. Each server has $k + 1$ ports, numbered from level-0 to level-k. Hence, BCube$_k$ has $N = n^{k+1}$ servers. Each level having n^k n-port switches. The construction of a BCube$_k$ is as follows: the n BCube$_{k-1}$s are numbered from 0 to $n - 1$ and the servers in each BCube$_{k-1}$ are numbered from 0 to $n^k - 1$. Then each level-k port of the ith server ($i \in [0, n^k - 1]$) in the jth BCube$_{k-1}$ ($j \in [0, n - 1]$) is connected to the jth port of the ith level-k switch. The BCube construction guarantees that switches only connect to servers and never connect directly to other switches, thus multipathing between switches is impossible. It is worth noting that this kind of architecture requires virtual bridging in containers to operate. Figure 1.3 shows an example of a BCube$_1$, with $n = 4$.

Similarly to BCube, DCell Guo et al. (2008a) uses servers equipped with multiple network ports and miniswitches to construct its recursive architecture. In DCell, a server is connected to several other servers and a miniswitch. Generally, a high-level DCell is constructed from low-level DCells. The connection between different DCell networks is typically done by using virtual bridging in containers. A DCell$_k$ ($k \geq 0$) is used to denote a level-k DCell. DCell$_0$ is the building block to construct larger DCells. It has n servers and a miniswitch ($n = 4$ for DCell$_0$ in Figure 1.4). All servers in DCell$_0$ are connected to the miniswitch.

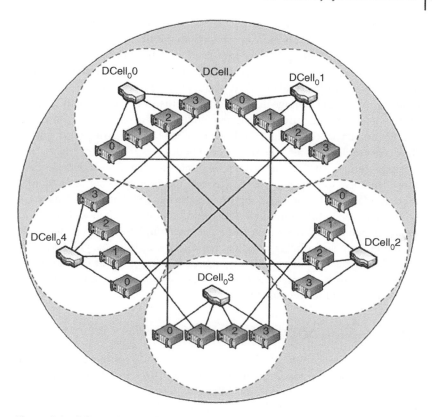

Figure 1.4 $DCell_1$ with $n = 4$.

In $DCell_1$, each $DCell_0$ is connected to all the other $DCell_0$s with one link; Figure 1.4 shows a $DCell_1$ example. $DCell_1$ has $n + 1 = 5$ $DCell_0$s. DCell connects the five $DCell_0$s as follows. It assigns each server a two-tuple $[a_1, a_0]$, where a_1 and a_0 are the level-1 and level-0 IDs, respectively. Thus, a_1 and a_0 take values from $[0, 5)$ and $[0, 4)$, respectively. Then two servers with two-tuples $[i, j - 1]$ and $[j, i]$ are connected with a link for every i and every $j > i$.

Each server has two links in $DCell_1$. One connects to its miniswitch, and hence to other nodes within its own $DCell_0$. The other connects to a server in another $DCell_0$. In $DCell_1$, each $DCell_0$, if treated as a virtual node, is fully connected with every other virtual node to form a complete graph. Moreover, since each $DCell_0$ has n inter-$DCell_0$ links, a $DCell_1$ can only have $n + 1$ $DCell_0$s, as illustrated in Figure 1.4. A $DCell_k$, is constructed in the same way to the above DCell$_1$ construction. The recursive DCell construction procedure (Guo et al., 2008a) is more complex than the BCube procedure.

To summarize, Fat-Tree topology originated in order to reduce core nodes bottleneck, by the fact that this topology grows horizontally. BCube and DCell architecture also called MDC offer the possibility of growing easily without huge change done by the administrators. Bcube can be seen as two-layer architecture with access nodes and core nodes, where the aggregation layer is dropped. DCell topology can be seen as one layer, where the access nodes play also the role of core nodes.

Server-centric DCN architectures, leading on recursive network structures, succeed to alleviate the bottleneck in core-layer switches thanks to redundant paths provided between servers. The entire DC fabric is built on servers while minimizing the set of deployed switches. Therefore, maintenance and management tasks become simpler. Moreover, network functions such as traffic aggregation, packet forwarding, etc., are delegated to servers. However, due to their recursive structure, server-centric structures significantly increase the number of servers, which would drastically increase the cabling complexity.

1.1.4 Enhanced DCN Architectures Overview

Despite the use of multi-gigabytes wired links and multiport switches in order to balance the load, the aforementioned DCN architectures are still facing flexibility and congestion challenges. Recently, a promising solution has investigated the possibility of augmenting the wired infrastructure by novel networking techniques, to enhance the capacity of DCNs. In the literature, the augmentation of such a DCN can mainly be achieved using two ways: (i) **optical** or (ii) **wireless** devices.

1.1.4.1 Optical DCN Architecture

Optical data center network (O-DCN) is a DCN architecture based on optical cabling and switching. Indeed, it has been found out that deploying such optical devices in DCs achieves a gain of 75% in IT power. First, on-demand high-speed links can be easily established thanks to the flexibility of optical network compared to the traditional wired DCN. Second, optical devices are able to ensure high bandwidth over longer ranges, and avoid, hence, the cost required for cabling along large distances. Further, O-DCNs deploy optical switches with high-radix ports, characterized by a low temperature, so as to reduce refrigeration cost. O-DCN can be classified in two main classes: (i) full optical DCN (all O-DCN) and (ii) hybrid optical DCN (hybrid O-DCN), detailed hereafter.

Full O-DCN Architectures In such architectures, all the control and data planes devices are optical. The key idea behind this full optical deployment is to provide high-speed bandwidth in the DCN. In this regard, O-DCN makes use of several techniques. First, optical circuit switching (OCS) (Chen et al., 2013) has been deployed in order to offer large bandwidth at the core layer. To do so, OCS DCN (Kachris and Tomkos, 2012) proceeds to preconfiguring the static routing paths in the switches. Second, optical packet switching (OPS), proposed in Chen et al. (2013), provides on-demand bandwidth in the DCN. In Ye et al. (2010), the datacenter optical switch (**DOS**) scalable DCN architecture has been propounded based on OPS technique. However, such an architecture suffers from low scalability. In addition, the elastic optical network (**EON**) (Talebia et al., 2014), is a kind of full O-DCN offering centralized on-demand flexibility in bandwidth switching.

Hybrid O-DCN Architectures Hybrid optical DCNs augment the wired DCNs by optical devices so that to provide extra bandwidth in an on-demand way by switching the connections in order to alleviate routing hop-counts. In doing so, hybrid O-DCNs succeed to minimize congestion effects on top of racks and to reduce traffic complexity by ensuring on-demand connections.

In this context, the authors of Chen et al. (2014) introduced a novel optical switching architecture (**OSA**) based on some techniques. Specifically, OSA makes use of a shortest path routing scheme and optical hop-to-hop switching in order to enable connectivity in DCN.

Moreover, **Helios** in Farrington et al. (2010), is a hybrid electrical–optical DCN, where each ToR is connected simultaneously to an electrical and an optical network. While electrical network is a Fat-Tree hierarchical structure, the optical one maintains a single optical connection on each ToR, with unlimited capacity. Helios deploys mirrors on a microelectro mechanical system to route the optical signals so as to alleviate traffic congestion at core level.

An additional example of hybrid O-DCN is **c-Through** Wang et al. (2010), a platform that includes a control and a data plane. The control plane measures an estimation of interrack traffic demands, then it dynamically calibrates circuits in a way that accommodates the new incoming flows. On the other side, the data plane isolates the electrical network from the optical one, and dynamically switches traffic from servers or ToRs onto the circuit or packet path. c-Through favors the use of optical paths as long as they are available, compared to the electrical routes. Nevertheless, it is worth pointing out that both of Helios and c-Through architectures fail to alleviate routing overheads.

FireFly is a wireless optical DCN architecture based on free-space optics (FSO) (Hamedazimi et al., 2014). The main advantage of such a design is that it provides a high data rate (≈ tens of Gbps) for long communication range while using low transmission power without interference. Specifically, servers in different racks communicate with each other using FSO reflected on ceiling mirrors.

Discussion In conclusion, enhancing DCN with optical technique succeed to satisfy many Cloud computing requirements. Particularly, it provides high-speed traffic with low-power consumption. Optical links alleviate the overhead compared to electric links. The aforementioned research optical approaches offer flexible switching solutions in order to make easy the bandwidth management for on-demand Cloud services. However, these designs still suffer from several limitations. First, O-DCN induces switching overhead. In fact, it requires the deployment of some modulation schemes in order to properly adjust bandwidth while switching connections, which is a challenging task. Second, O-DCN cannot be deployed in large-scale environments so far because of the high cost of optical transceivers and their long latency. Third, a significant reconfiguration latency of roughly 10 ms is induced by O-DCN which would affect applications quality of service (QoS), such as online services.

1.1.4.2 Wireless DCN Architecture

To address the challenges of both wired and optical DCN in terms of cabling complexity, deployment cost, scalability, and so on, Wireless DCN (W-DCN) has been recently explored. W-DCN architecture deploys wireless antennas, operating in the 60 GHz frequency band, to connect pairs of ToR switches. In doing so, the wired infrastructure is augmented with interrack wireless links. The main insight behind this approach is to investigate the high data transfer rate of this new emerging technique, that can reach 7 Gbps, in order to enhance DCN performance. Actually, a 60 GHz wireless link makes use of the physical beamforming technique so that the transmitted signal is concentrated in a specific direction enhancing while mitigating interference. The related wireless DCN architectures found in the literature could be classified into: (i) hybrid W-DCN and (ii) full W-DCN. Hereafter, we will detail the most relevant wireless DCN architectures.

Hybrid Wireless DCN Architectures In such an architecture, both wired and wireless infrastructures are used in the same DCN. Wireless augmentation of DCN has been first explored by the authors of Ramachandran et al. (2008)

in order to reduce cabling complexity in the wired DCN while enhancing network flexibility. The main idea behind their design is to replace some of wired bottleneck links by wireless connections operating in the 60 GHz range. Besides, Vardhan et al. (2010) designs a wireless DCN based on IEEE 802.5.3c standard (IEEE Std 802.15.3c-2009, 2009) in the wireless 60 GHz communications. To study the feasibility of such technique in DCN, the authors emulate three-tier and Fat-Tree architectures with wireless links. To do so, they propose node placement algorithms to assign nodes to racks.

Later on, **Flyway-based** DCN architecture (Greenberg et al., 2009b, Kandula et al., 2009) has been propounded in order to alleviate congestion on hotspot links in the VL2 architecture (Greenberg et al., 2009a). However, Flyway links are created on-demand in the DCN as long as there is congestion on the ToR and struggle to meet all the challenges of DCN such as scalability, high traffic load, and interference.

The authors of Cui et al. (2011c) have proposed a hybrid wired/wireless DCN architecture, where each ToR, considered as a wireless transmission unit (WTU), is equipped with a set of wireless 60 GHz radios. This hybrid architecture investigates the use of wireless infrastructure in order to reduce the congestion level of congested nodes and to handle unbalanced traffic demands in DCN.

In Cui et al. (2011a), the authors envision a hybrid Ethernet/wireless three-layered DCN architecture. Congestion on core layer is alleviated by deploying 60 GHz wireless antennas on top of racks, without needing to rearrange servers in the same rack.

To further enhance the DCN performance, some research work papers have investigated the use of beamforming technique while designing hybrid DCN architectures. Particularly, 3D beamforming has been presented in Zhang et al. (2011) and Zhou et al. (2012) in order to boost the transmission range and 60 GHz spectrum reuse in DCNs. Basically, the enhanced design sets up indirect LOS path by making use of ceiling reflectors. These enable the interconnection of wireless antennas that are not placed in the same transmission range. Typically, the horn antenna placed on each sending rack radiates the signal in some points on the reflector, and the latter transmits the signal to the receiver. In doing so, obstacles are eliminated and racks could communicate directly in one hop. While this 3D beamforming architecture significantly extends wireless coverage distance, it requires the absence of obstacles between the top of rack/container and the ceiling which is not guaranteed in real DC environments.

The authors of Katayama et al. (2011) investigate the use of steered-beam antennas in order to build a robust wireless **crossbar** switch-centric DCN architecture. In such a design, wired cabling is used only for intrarack links or to interconnect racks within the same row. On the other hand, wireless steered-beam antennas are deployed on adjacent ToRs while constituting a wireless crossbar so that cabling task is simplified and installation cost is reduced.

Angora architecture recently proposed in Zhu et al. (2014) propounds a robust wireless topology for the control plane while data is completely transiting over wired infrastructure. To do so, 3D beamforming radios are deployed on racks based on Kautz graphs, so that network latency is reduced by minimizing the path length between communicating racks. Moreover, Angora alleviates interflow interference by statically calibrating the directions of the deployed horn/array antennas. Unfortunately, the static 3D direction of antennas may strongly limits the usage of spectrum.

In Li et al. (2014), a **spherical mesh** is a wireless DCN where racks within the same transmission range are regrouped into a spherical unit. The main idea is to take profit of the geometric characteristics of the spheres to eliminate link congestion by placing antennas over them. Moreover, the spherical mesh DCN reduces the network diameter by dividing the DCN into several units.

RUSH DCN architecture is proposed in Han et al. (2015), which is a hybrid DCN based on the common three-layer tree topology. In RUSH, each ToR is equipped by only one directional 60 GHz antenna and wireless interrack links are used to minimize congestion. For that end, the authors propose a scheduling framework to jointly route flows and schedule wireless antennas.

In Cui et al. (2011b), **Diamond** DCN architecture is improved by deploying 3D wireless rings. Unlike common hybrid designs, Diamond is a hybrid wired/wireless DCN where all links between servers are wireless, whereas links connecting servers to ToRs or connecting ToRs are wired. The rings consist in regular polygons which are constructed by racks and metal reflectors, while the layers contain the servers inside racks belonging to the same level. The main reason behind the use of 3D ring reflection spaces (RRSs) is their low-cost and their ability to provide wireless links by multireflection of signals over metal. Diamond feasibility has been studied based on a real testbed.

VLCcube is a hybrid DCN architecture which is propounded in Luo et al. (2016). It is an augmented Fat-Tree structure that specifically organizes all racks into a wireless Torus structure while making use of the visible light communications (VLC) technique to generate high-speed links. In doing so, all racks are connected based on VLC links. VLC is a promising solution that guarantees low cost and important bandwidth. Moreover, VLC links do not require mechanical or electronic control.

Full Wireless DCN Architectures A completely wireless DCN architecture has been propounded in Shin et al. (2013), based on a Cayley graph, thereby named Cayley Data Center structure (**Cayley DC**). The servers are grouped into cylindrical racks. Each one is composed of five levels named stories. A story consists of 20 containers of servers. Racks are attached to densely wireless connected mesh topology with the aim of maximizing the number of active wireless links. Specifically, the Cayley DC uses wireless links not only for interrack communications but also inside racks, thanks to the mesh structure. In order to alleviate interference effects, this strategy makes use of beamforming technique with fixed-direction antennas.

Discussion To summarize, most of the relevant research work published in the recent years approves the feasibility and the efficiency of deploying 60 GHz wireless technology as an extension of conventional wired DCN architectures. Hybrid wireless/wired DCN have proven a significant capability to enhance network performance and to address the major data center issues, namely scalability, flexibility, and cabling complexity.

1.2 Comparison Between DCN Architectures

In this section, we will present a qualitative comparison between the reviewed DCN architectures while considering some specific criteria: scalability, bandwidth, cabling complexity, deployment cost, and fault tolerance. Scalability refers to the ability of the proposed architecture to easily scale and deploy more devices. Bandwidth represents the proportion of available bandwidth between servers and switches, while cabling complexity refers to the multitude of cables in the DCN induced by link redundancy. The overheads and the cost of deployment in DCN are also crucial factors that refer to the number of switches and links and their corresponding construction and deployment cost. Finally, fault-tolerance defines the ability of the designed architecture to deal with switch and link failures.

Table 1.1 illustrate a comparison between different DCN architectures based on the aforementioned aspects.

1.3 Proposed HDCN Architecture

We envision a hybrid (wireless/wired) data center network (HDCN) architecture built over a three-stage CLOS topology. Indeed, as explained

Table 1.1 Summary and analysis of DCN architectures

Architecture	Technique	Scale	Bandwidth	Scalability	Cabling complexity	Cost	Fault-tolerance
Tree-based Chen et al. (2016)	Wired	Small	Low	Bad	High	High	Bad
CLOS Al-Fares et al. (2008)	Wired	Medium	Medium	Medium	High	High	Medium
FatTree Al-Fares et al. (2008)	Wired	Medium	Medium	Medium	High	High	Medium
ElasticTree Heller et al. (2010)	Wired	Medium	Medium	Medium	High	High	Medium
PortLand Mysore et al. (2009)	Wired	Medium	Quite high	Medium	High	High	Good
Diamond Sun et al. (2014)	Wired	Medium	High	Medium	High	Low	Medium
VL2 Greenberg et al. (2009a)	Wired	Large	Quite high	Medium	High	High	Medium
Monsoon Greenberg et al. (2008)	Wired	Large	Quite high	Medium	High	High	Medium
FBFLY Abts et al. (2010)	Wired	Large	High	Medium	High	High	Medium
FlaNet Lin et al. (2012)	Wired	Large	High	Low	High	High	Medium
C-FBFLY Csernai et al. (2015)	Wired	Large	High	Low	High	High	Medium

Approach	Reference							
DCell	Guo et al. (2008b)	Wired	Large	High	Good	High	Medium	Good
FiConn	Li et al. (2009)	Wired	Large	High	Good	Medium	Medium	Good
O-DCN	Wang et al. (2010) / Farrington et al. (2010) / Hamedazimi et al. (2014)	Optical	Small	Very high	Medium	High	High	Bad
BCube	Guo et al. (2009a)	Wired	Small	Very high	Good	Medium	Medium	Very good
CamCube	Abu-Libdeh et al. (2010)	Wired	Large	High	Good	Very high	High	Good
Flyway-based	Kandula et al. (2009) / Greenberg et al. (2009b)	60 GHz/Ethernet	Medium	Very high	Good	Medium	Medium	Good
Wireless Fat-Tree	Vardhan et al. (2010)	60 GHz/Ethernet	Medium	Very high	Good	Medium	Medium	Medium
Hybrid DCN	Cui et al. (2011c)	60 GHz/Ethernet	Medium	Very high	Good	Medium	Medium	Medium
Hybrid DCN	Cui et al. (2011a)	60 GHz/Ethernet	Medium	Very high	Good	Medium	Medium	Medium

Table 1.1 (Continued)

Architecture	Technique	Scale	Bandwidth	Scalability	Cabling complexity	Cost	Fault-tolerance
3D Beamforming Zhang et al. (2011) Zhou et al. (2012)	3D Beamforming	Medium	Very high	Good	Medium	High	Medium
Wireless crossbar Katayama et al. (2011)	60 GHz	Medium	Very high	Good	Medium	High	Medium
Cayley DC Shin et al. (2013)	60 GHz	Medium	Very high	Good	Medium	High	Good
Angora Zhu et al. (2014)	60 GHz/ Ethernet	Medium	Very high	Good	Medium	High	Medium
Spherical mesh Li et al. (2014)	60 GHz/ Ethernet	Medium	Very high	Good	High	High	Medium
RUSH Han et al. (2015)	60 GHz/ Ethernet	Medium	Very high	Good	Medium	High	Medium
3D Diamond Cui et al. (2011b)	3D Beamforming/ wired	Medium	Very high	Good	Medium	High	Medium
VLCcube Luo et al. (2016)	VLC	Medium	Very high	Good	Medium	High	Medium

in Section 1.1, CLOS-based architecture has been widely considered in modern DCs and has proven a high performance and resiliency.

To mimic a real data center environment, our CLOS-based HDCN architecture follows the CISCO's massively data center (MSDC) model (CISCO Systems, 2014). In fact, MSDC is a promising framework capable of supporting huge volume of traffic. To augment the wired infrastructure in HDCN by wireless links, we make use of 60 GHz wireless technology. In doing so, traffic can be forwarded over wireless and/or wired links which will alleviate the congestion load and hence improve the network performance.

In this section, we will first highlight the main properties of MSDC model. Second, we will focus on the wireless infrastructure in the HDCN by presenting the: (i) 60 GHz technology, (ii) IEEE 802.11ad standard, and (iii) deployed beamforming mechanism.

1.3.1 HDCN Architecture Based on MSDC Model

CISCO's massively scalable data center (MSDC) is a framework model that has been widely used by data center architects to build flexible DCs supporting applications distributed across thousands of servers.

Typically, MSDC is built based on a CLOS-based topology with a short spine layer serving as the aggregation switches, and a long leaf layer serving as the access layer. Specifically, a three-stage CLOS MSDC architecture using 32 port switches, and can thus connect up to 8192 servers. Based on the CISCO's MSDC reference CISCO Systems (2014), our HDCN architecture follows a three-stage CLOS topology formed by: (i) spine layer using Nexus 7000 switches and (ii) leaf layer deploying Nexus 3000 platform.

Each leaf connects to all spines. In doing so, our MSDC-based HDCN network provides multiple paths for interrack communications between servers. To leverage the multiple paths available between leaf and spine switches, MSDC data center deploys both OSPF routing and equal cost multipathing (ECMP) protocols. ECMP maximizes the load balancing of wired links' usage by dividing the traffic through multiple equal cost routes. Hereafter, we will detail the load-balancing ECMP mechanism used in our HDCN.

1.3.1.1 ECMP Protocol

ECMP (Network Working Group – Request for Comments: 2992, 2000) is the most commonly used protocol in today's data centers, for the traffic load balancing across redundant shortest routing paths. The main idea of ECMP is to divide the traffic through multiple equal cost routes. Basically, this technique is a selection tool that finds the convenient route for each transmitted packet and this by choosing the next hop from the computed OSPF routes. Mainly, two modes of load balancing are associated with ECMP: (i) per packet mode, where the packets of the same flow may have different routes and (ii) per flow mode, where the packets of the same flow are forwarded to the same next-hop, ensuring the ordered arrival of packets in transmission control protocol (TCP) mode.

In this chapter, we generate traffic, in HDCN, based on User Datagram Protocol (UDP). Consequently, based on ECMP requests for comments (RFC) (Network Working Group – Request for Comments: 2992, 2000), ECMP activates (i) the mode per-packet to maximize the load balancing and (ii) Round Robin scheduler to select the next hop (outgoing interface) for each packet.

1.3.2   60 GHz Technology in HDCN

As in prior work (Halperin et al., 2011, Kandula et al., 2009, Ramachandran et al., 2008, Shin et al., 2013), we propose in this chapter to deploy 60 GHz wireless technique in order to enhance our hybrid DCN architecture. Specifically, wireless infrastructure in our HDCN is based on IEEE 802.11ad. This standard, presented by the working group TGad as the enabler of next generation Multi-Gbps WiFi, takes advantages of available spectrum in the unlicensed 57–66 GHz band. It offers four orthogonal physical channels whose center frequencies are respectively fixed at 58.32, 60.48, 62.64, and 64.8 GHz. The capacity of each wireless channel reaches 6.7 Gbps over a short range. Consequently, the whole network of a data center can be seen as personal basic service set (PBSS). Indeed, PBSS is IEEE 802.11ad wireless LAN in which stations communicate directly with each other (i.e. ad hoc network, no need of access point) (IEEE Std 802.11ad 2012, 2012). Note that each node in PBSS is denoted by directional multi-gigabit station (DMG-STA). The latter is defined in the standard as a station operating at a frequency above 45 GHz and can support a throughput greater or equal to 1 Gbps.

In PBSS network, one DMG-STA must assume the role of controller and is denoted by PBSS control point (PCP). It ensures the QoS traffic scheduling, resource allocation, control admission, association/disassociation, etc. In other words, the PCP has a global view of nodes in PBSS. PCP is a global controller responsible for the (i) management of the Hybrid DCN and (ii) optimization of the resource usage and flows forwarding. It is worth noting that the communication between the PCP and all the DMG-STA in PBSS should be ensured over wireless network. However, some WTU deployed over DMG-STA cannot reach the PCP in wireless one-hop. In our architecture, we propose that communications between the PCP and WTUs will be supported by the wired infrastructure (e.g. Ethernet, OpenFlow). In doing so, we can see our architecture as a Software-Defined Network. In fact, the control plane is centralized in the PCP and WTUs support only the data plane (i.e. transmission of frames). IEEE 802.11ad standard defines three frame classes. In our Hybrid DCN (i.e. PBSS), we leverage the frames of Class 1. The latter contains three frame types: (i) control frames, (ii) data frames, and (iii) management frames. Concerning the control frames, we only make use of acknowledgement frames. They are transmitted over a single carrier modulation by setting the modulation and coding scheme (MCS) to 0. The latter corresponds to differential binary phase-shift keying (DBPSK) modulation, code rate is $\frac{1}{2}$, data rate is 27.5 Mbps, and receiver sensitivity is −78dBm. On the other hand, data frames are transmitted over orthogonal frequency-division multiplexing (OFDM) modulation by setting MCS to 24. The latter corresponds to 64-QAM modulation, code rate is $\frac{13}{16}$, data rate is 6756.75 Mbps (i.e. maximum data rate), and receiver sensitivity is −47dBm. Finally, the management frames are transmitted over the wired infrastructure.

Based on this specification, we propose to deploy at each ToR, a WTU composed of a set of four directional transceivers/antennas. Each transceiver is, hence, assigned to one wireless channel. Note that the four transceivers in WTU are independent, due channel orthogonality, and can be simultaneously exploited. In doing so, any rack in the data center can communicate over the wired ports (i.e. ToR) and/or using wireless channels.

It is worth pointing out that the wireless 60 GHz communication is faced with several challenges due to the free space propagation loss. The latter is due to the low-power density, and results in a short transmission range. Moreover, wireless links are prone to interfere in HDCN environment which deeply affects transmission stability. To address these limitations, we explore in this thesis beamforming technique so that to minimize the propagation loss and increase coverage distance.

1.3.3 Beamforming Technique in HDCN

The beamforming is a physical layer technique that concentrates transmission power in a specific direction (i.e. beam) so that the link rate is enhanced.

Unlike omnidirectional antennas radiating signal in a uniform way (circle), smart directional transceivers are capable of transmitting signal in one single beam (angle) by targeting only the direction of the destination. Typically, a directional antenna is in general composed by (i) an array of

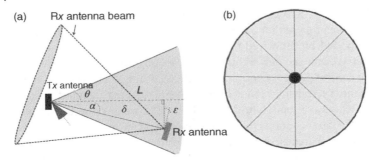

Figure 1.5 Switched-beam antenna model: (a) spherical coordinate system and (b) beams.

antenna elements (beams) and (ii) a signal processor adjusting the radiation of the latter.

Mainly, current 60 GHz beamforming antennas are available either as horn antennas (Halperin et al., 2011), phased-array antennas, or switched-beam antennas (Zhu et al., 2014). While the phased-array transceivers are steerable devices that appropriately steer each beam at the desired target direction, horn antennas are in general used for fixed links, in long-range outdoor environments. Recent researches Zhu et al. (2014), Zhou et al. (2012) claim that both array and horn antennas require a mechanical rotation mechanism at each single communication to adjust the beam direction. This frequent antenna rotation induces an extra delay estimated to equal 50 ns for array antennas and to range from 0.01 to 1 second for horn antennae (Zhou et al., 2012). Based on these observations and as recommended by Zhu et al. (2014), we deploy, in this thesis, switched beam antennas to avoid performance degradation. In fact, such devices have been considered to be less complex than the other smart radios and are cheaply implemented. As depicted in Figure 1.5b, a switched beam antennae is characterized by an array of N beams (i.e. sectors). Each one covers an angle of $2\Pi/N$. Accordingly, the transmitting antenna switches to (i.e. selects) the beam achieving the highest gain while covering the destination. The receiving antenna senses the signal on all the sectors and exploits only the one achieving the maximum gain. The signal coming from potential interfering antennas is either not received or significantly weak.

We assume the geometric signal propagation model (Shin et al., 2013) based on a spherical coordinate system with origin the transmitting antenna as shown in Figure 1.5a. The receiver antenna is characterized by radius δ, azimuth θ as shown in Figure 1.5a. Note that we assume 2D beamforming and hence elevation is equal to 0.

Our HDCN architecture is illustrated in Figure 1.6.

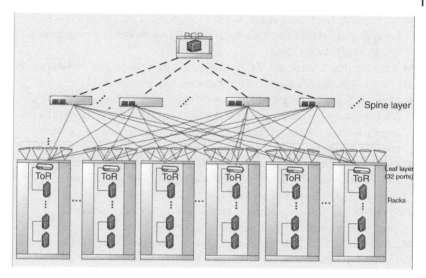

Figure 1.6 Hybrid CISCO MSDC architecture of a DCN.

1.4 Conclusion

In this chapter, we provided an overview of DCN architectures. First, we proposed a taxonomy classifying the relevant DCN structures into three main classes: (i) switch-centric, (ii) server-centric, and (iii) enhanced DCN architectures. We deeply analyzed the key properties of each class. Afterward, we provided a qualitative comparison study between the different DCN architectures. Finally, we presented our chosen hybrid DCN architecture based on (i) Cisco's MSDC framework and (ii) wireless 60 GHz technique. In the next chapter, we will present a detailed review on the most relevant research strategies in the literature tackling wireless/wired resource allocation problem for both one-hop and multihop communications in HDCN.

References

D. Abts, M.R. Marty, P.M. Wells, P. Klausler, and H. Liu. Energy proportional datacenter networks. *ACM SIGARCH Computer Architecture*, June 2010. 38 338–347.

H. Abu-Libdeh, P. Costa, A. Rowstron, G. OShea, and A. Donnelly. Symbiotic routing in future data centers. *ACM Special Interest Group on Data Communication (SIGCOMM)*, August 2010.

M. Al-Fares, A. Loukissas, and A. Vahdat. A scalable, commodity data center network architecture. *ACM Special Interest Group on Data Communication (SIGCOMM)*, August 2008.

M. Chen, H. Jin, Y. Wen, and V.C. Leung. Enabling technologies for future data center networking: a primer. *IEEE Network*, July 2013. 27 8–15.

K. Chen, A. Singla A. Singh, K. Ramachandran, L. Xu, Y. Zhang, X. Wen, and Y. Chen. OSA: an optical switching architecture for data center networks with unprecedented flexibility. *IEEE/ACM Transactions on Networking*, April 2014. 22 498–511.

T. Chen, X. Gao, and G. Chen. The features, hardware, and architectures of data center networks: a survey. *Journal of Parallel and Distributed Computing*, October 2016. 96 45–74.

CISCO Systems. Massively scalable data center (MSDC) design and implementation guide. Technical report, October 2014.

M. Csernai, F. Ciucu, R. Braun, and A. Gulyas. Towards 48-fold cabling complexity reduction in large flattened butterfly networks. *IEEE INFOCOM*, May 2015. 109–111.

Y. Cui, H. Wang, X. Cheng, and B. Chen. Wireless data center networking. *IEEE Wireless Communications*, December 2011a. 18 46–53.

Y. Cui, S. Xiao, X. Wang, Z. Yang, C. Zhu, X. Li, L. Yang, and N. Ge. Diamond: nesting the data center network with wireless rings in 3D space. *USENIX NSDI*, March 2011b.

Y. Cui, H. Wang, and X. Cheng. Channel allocation in wireless data center networks. *IEEE International Conference on Computer Communications (INFOCOM)*, April 2011c.

N. Farrington, G. Porter, S. Radhakrishnan, H.H. Bazzaz, V. Subramanya, Y. Fainman, G. Papen, and A. Vahdat. Helios: a hybrid electrical/optical switch architecture for modular data centers. *ACM SIGCOMM*, August 2010.

A. Greenberg, J. Hamilton, D. Maltz, and P. Patel. The cost of a cloud: research problems in data center networks. *ACM Special Interest Group on Data Communication (SIGCOMM)*, August 2008.

A. Greenberg, A. Greenberg, J.R. Hamilton, N. Jain, S. Kandula, C. Kim, and P. Lahiri. VL2: a scalable and flexible data center network. *ACM Special Interest Group on Data Communication (SIGCOMM)*, August 2009a.

A. Greenberg, N. Jain, S. Kandula, C. Kim, P. Lahiri, D. Maltz, P. Patel, and S. Sengupta. VL2: a scalable and flexible data center network. *ACM SIGCOMM*, August 2009b.

C. Guo, H. Wu, K. Tan, L. Shi, Y. Zhang, and S. Lu. DCell: a scalable and fault-tolerant network structure for data centers. 38 (4): 75–86, 2008a.

C. Guo, H. Wu, K. Tan, L. Shi, Y. Zhang, and S. Lu. DCell: a scalable and fault-tolerant network structure for data centers. *ACM SIGCOMM*, August 2008b.

C. Guo, G. Lu, D. Li, H. Wu, X. Zhang, Y. Shi, C. Tian, Y. Zhang, and S. Lu. BCube: a high performance, server-centric network architecture for modular data centers. *ACM SIGCOMM*, August 2009a. 39 63–74.

C. Guo, G. Lu, D. Li, H. Wu, X. Zhang, Y. Shi, C. Tian, Y. Zhang, and S. Lu. BCube: a high performance, server-centric network architecture for modular data centers. *ACM SIGCOMM Computer Communication Review*, 2009b.

D. Halperin, S. Kandula, J. Padhye, P. Bahl, and D. Wetherall. Augmenting data center networks with multi-gigabit wireless links. *ACM Special Interest Group on Data Communication (SIGCOMM)*, August 2011.

N. Hamedazimi, Z. Qazi, H. Gupta, V. Sekar, S.R. Das, J.P. Longtin, H. Shah, and A. Tanwer. FireFly: a reconfigurable wireless data center fabric using free-space optics. *ACM Special Interest Group on Data Communication (SIGCOMM)*, August 2014.

K. Han, Z. Hu, J. Luo, and L. Xiang. RUSH: routing and scheduling for hybrid data center networks. *IEEE Conference on Computer Communications (INFOCOM)*, August 2015.

B. Heller, S. Seetharaman, P. Mahadevan, Y. Yiakoumis, P. Sharma, S. Banerjee, and N. McKeown. ElasticTree: saving energy in data center networks. *The 7th USENIX Conference on Networked Systems Design and Implementation*, April 2010.

IEEE Std 802.15.3c-2009. IEEE Standard for information technology – telecommunications and information exchange between systems – local and metropolitan area networks – specific requirements. Part 15.3: Wireless Medium Access Control (MAC) and Physical Layer (PHY) Specifications for high rate wireless personal area networks (WPANs) amendment 2: Millimeter-wave-based alternative physical layer extension. *(Amendment to IEEE Std 802.15.3-2003)*, December 2009.

IEEE Std 802.11ad 2012. IEEE Standard for Information technology-Telecommunications and information exchange between systems-Local and metropolitan area networks-Specific requirements-Part 11: Wireless LAN Medium Access Control (MAC) and physical layer (PHY) Specifications Amendment 3: Enhancements for Very High Throughput in the 60 GHz Band. *IEEE Std 802.11ad 2012*, December 2012.

C. Kachris and I. Tomkos. A survey on optical interconnects for data centers. *IEEE Communications Surveys and Tutorials*, January 2012. 14 1021–1036.

S. Kandula, J. Padhye, and P. Bahl. Flyways to de-congest data center networks. *ACM SIGCOMM HotNets*, August 2009.

Y. Katayama, K. Takano, Y. Kohda, N. Ohba, and D. Nakano. Wireless data center networking with steered-beam mmWave links. *IEEE WCNC*, April 2011.

D. Li, C. Guo, H. Wu, K. Tan, Y. Zhang, and S. Lu. FiConn: using backup port for server interconnection in data centers. *IEEE Infocom*, May 2009.

Y. Li, F. Wu, X. Gao, and G. Chen. SphericalMesh: a novel and flexible network topology for 60 GHz-based wireless data centers. *IEEE/CIC International Conference on Communications (ICCC)*, October 2014.

D. Lin, Y. Liu, M. Hamdi, and J. Muppala. FlatNet: towards a flatter data center network. *IEEE GlobeCom*, December 2012.

L. Luo, D. Guo, J. Wu, S. Rajbhandari, T. Chen, and X. Luo. VLCcube: a VLC enabled hybrid network structure for data centers. *IEEE Transactions on Parallel and Distributed Systems*, December 2016. 28 2088–2102.

R. Mysore, A. Pamboris, N. Farrington, N. Huang, P. Miri, S. Radhakrishnan, V. Subramanya, and A. Vahdat. PortLand: a scalable fault-tolerant layer 2 data center network fabric. *ACM Special Interest Group on Data Communication (SIGCOMM)*, August 2009.

IETF, RFC 2992. Analysis of an equal-cost multi-path algorithm. Technical report, November 2000.

K. Ramachandran, R. Kokku, R. Mahindra, and S. Rangarajan. 60 GHz data-center networking: wireless ==¿ worry less? Technical report, NEC Laboratories America, July 2008.

J.Y. Shin, E.G. Sirer, H. Weatherspoon, and D. Kirovski. On the feasibility of completely wireless datacenters. *IEEE/ACM Transactions on Networking (TON)*, October 2013. 21 1666–1679.

Y. Sun, J. Chen, Q. Liu, and W. Fang. Diamond: an improved fat-tree architecture for largescale data centers. *Journal of Communications*, January 2014. 9 91–98.

S. Talebia, F. Alamb, I. Katibb, M. Khamisb, R. Salamab, and G.N. Rouskas. Spectrum management techniques for elastic optical networks: a survey. *Optical Switching and Networking*, July 2014. 13 34–48.

H. Vardhan, N. Thomas, S.R. Ryu, B. Banerjee, and R. Prakash. Wireless data center with millimeter wave network. *IEEE GlobeCom*, December 2010.

G. Wang, D.G. Andersen, M. Kaminsky, K. Papagiannaki, T.S. Eugene, M. Kozuch, and M. Ryan. c-Through: part-time optics in data centers. *ACM Special Interest Group on Data Communication (SIGCOMM)*, October 2010.

X. Ye, Y. Yin, S.J.B. Yoo, P. Mejia, R. Proietti, and V. Akella. DOS: a scalable optical switch for datacenters. *6th ACM/IEEE Symposium on Architectures for Networking and Communications Systems (ANCS)*, October 2010.

W. Zhang, X. Zhou, L. Yang, Z. Zhang, B.Y. Zhao, and H. Zheng. 3D beamforming for wireless data centers. *Hotnets*, November 2011.

X. Zhou, Z. Zhang, Y. Zhu, Y. Li, S. Kumar, A. Vahdat, B.Y. Zhao, and H. Zheng. Mirror mirror on the ceiling: flexible wireless links for data centers. *ACM SIGCOMM*, August 2012.

Y. Zhu, X. Zhou, Z. Zhang, L. Zhou, A. Vahdat, B.Y. Zhao, and H. Zheng. Cutting the cord: a robust wireless facilities network for data centers. *ACM Conference on Mobile Computing and Networking (MobiCom)*, September 2014.

2

Data Center Optimization Techniques

Dallal Belabed

Airbus Defense and Space, Airbus, Yvelines, Elancourt, France

Abstract

We synthetically provide in this section a brief overview about Data Center optimization techniques. In the first section, we first briefly introduce an overview of layer 2, layer 3, and up layers routing solutions. In the second section, we present useful works on virtual network embedding, and discuss state of the art energy efficiency consolidation, followed by an overview on Traffic Engineering (TE) techniques from link-state TE to introducing the transmission control protocol (TCP) fairness models.

2.1 Ethernet Switching and Routing

In the last decade, the evolution of the Data Center, metropolitan area network (MAN), and wide area network (WAN), has introduced new requirements. In an attempt to take advantage of Ethernet flexibility, all institutions such as Internet Engineering Task Force (IETF) (Bradner, 1999), and the Institute of Electrical and Electronics Engineers (IEEE) have introduced recent advanced standardizations in Ethernet technology for higher transmission rates and better topology utilization leading to better link utilization.

Hence, several evolutions in the legacy of Ethernet switching architecture in terms of traffic engineering (TE) features have occurred leading to the possibility of deploying Ethernet within the core of large-scale networks. In this section, we will provide the state of the Ethernet Switching and Routing protocols.

Management of Data Center Networks, First Edition. Edited by Nadjib Aitsaadi.
© 2021 The Institute of Electrical and Electronics Engineers, Inc.
Published 2021 by John Wiley & Sons, Inc.

Ethernet forwarding was originally designed for local area network (LAN), it was based on the 3F concepts: filter, forward, and flood. If the Media Access Control (MAC) destination of the incoming frame is on the same network or subnet than that of the port where the frame comes from, the switch does not forward the frame, dropping it. This is known as "filtering." If the destination is not on the same network, after looking at its MAC table, it will then forward the frame to the appropriate segment, or usually default gateway, called forwarding. If the MAC destination address is not in the table at all, it will then forward the frame out all of its ports except the port it was received on, this process is known as flooding.

In the beginning, the aim of forwarding protocols was to avoid loops on small networks. In this context, the spanning tree protocol (STP) (Perlman, 1985) or IEEE 802.1D, known as the mother of the Internet, was born. STP ensures the frames forwarding to avoid loops by giving a logical topology in the form of a tree, hence, the path from any source to any destination is unique and follows the tree. STP chooses a bridge from the topology as a root bridge and ensures that all the paths from the nodes to the root node are the shortest-path computed as the cumulative link cost path. Thus, two neighbored nodes on the physical topology using STP do not have any guarantee to be caring by the shortest path. That is why the choice of the root bridge dictates the efficiency of the resulting logical topology.

Mainly, STP goes through the following steps:

- Elects the root bridge, based upon static parameters, namely, the lowest bridge identifier (BID).
- Computation of the minimum cost path from each node to the root.
- Designated port election. For each network segment, choose the designated port, on which the bridge is responsible for forwarding data. As an example, see Figure 2.1, supposing that the cost of crossing of every network segment is 1, the shortest path from the network segment e will be carried through the switch 2. Consequently, the designated port for the network segment e is the port, which connects it to the switch 2.
- *Root port selection*: The port that gives the best path from a specific bridge to the root bridge. Supposing that the cost of crossing of every network segment is 1, the shortest path from switch 4 to the switch root passes by the network segment c. Consequently, the root port for the switch 4 is the one which leads to the network segment c.
- Blocked port, breaks loops by blocking the ports associated to a link that are not root nor designated ports.

Trivially, STP is not a scalable protocol, and its main drawbacks are deactivating some links, i.e. a nonefficient utilization of the throughput,

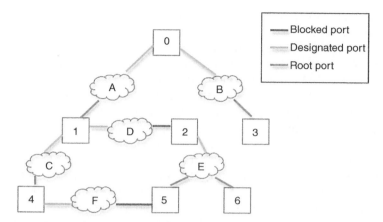

Figure 2.1 Spanning tree protocol.

no guarantee regarding the shortest path between any neighbor nodes, particularly when the physical topologies is big, after a topology change, STP convergence may take minutes, among others, this is due to the re-execution of the STP algorithm, to the period of Forward Delay + MaxAge (15 + 20 = 35), and to the configuration bridge protocol data units (BPDUs) that are only exchanged by the root bridge every Hello time (two seconds).

In 1998, as an answer to the STP time convergence, rapid spanning tree protocol (RSTP) has been proposed by the IEEE as IEEE 802.1w (Vojdak, 2008). The general functioning of RSTP is similar to STP. The main differences are the failure of the root bridge is detected by three hellos, that is six seconds with the default values; the introduction of an alternate port, which is an alternative port distinct to the root port, which consists of the alternative shortest path to the root bridge and changes directly in the forwarding state in case of the port root failure; and the backup port, i.e. in case of a redundant connection of a segment toward another bridge (two connections toward the same domain of collision). One of the ports will have the role of the Backup.

For the purpose of settling the unused links by the STP or RSTP protocols, the virtual local area network (VLAN) idea was introduced. VLAN can be defined as a logical topology upon the physical topology, where many VLANs can cohabitate together. There are different ways to define VLAN: it can be defined based on the network interface card (NIC) port identifier, called VLAN layer 1; it can be defined by MAC addresses (VLAN layer 2) or by IP addresses (VLAN layer 3), in the following text of this chapter, we are interested on the VLAN on layer 2, more specifically the tagged link where each frame is tagged.

The first standard of tagged frames was done by IEEE called 802.1Q, in 1998. IEEE 802.1Q (Thaler et al., 2013) is a successor of the Cisco inter-switch link (ISL) protocol. The standard IEEE 802.1Q has introduced the VLAN tag, which completes the Ethernet header (see Figure 2.2). In IEEE 802.1Q four additional bytes were added to the traditional Ethernet frame, presented in figure 1Q, two bytes represent the tag protocol identifier TPID (16 bits), fixed to 0x8100, and the two other bytes represent tag control information (TCI). The TCI is also divided into priority code point (PCP) three bits, drop eligible indicator (DEI) 1 bit, and VLAN identifier (VID) 12 bits, where the PCP refers to the standard IEEE 802.1p (EK Niclas, 1999). Since with three bits, we can code eight levels of priority from 0 to 7, these eight levels are used to fix a priority to the frames of one VLAN with regard to the other VLANs; DEI indicates if frames are eligible to be dropped in the presence of congestion; and finally the VID allows 4094 VLANs.

Others VLAN standards were introduced in order to increase the amount of VLANs, the most important are the provider bridges (PB) with IEEE 802.1ad QinQ (IEEE Std 802.1ad-2005, 2006) and provider backbone bridges (PBB) 802.1ah (Bridges, 2008) M-in-M.

The Q in Q protocol typically introduces S-TAG field, Q in Q data frame format is presented in Figure 2.3, the S-TAG provider bridges will differentiate between customers. This standard is commonly referred to "QinQ" or "VLAN" stacking, where C-tag represents the Customer VLAN and S-Tag represents Service provider VLAN. Mainly the ingress Provider Bridge stacks the server provider (SP) VLAN Id at the S-TAG. Inside the network provider, the Ethernet frames are forwarded using the customer MAC addresses and the VLAN ID using 3Fs, MAC tables, and STP. However, VLAN tag can only accommodate 4094 instances.

On the 802.1ah or M-in-M, essentially, PBB encapsulates an end-user's frame; it can be end customer or Client Service Provider; it encapsulates

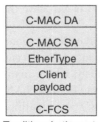
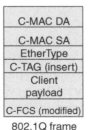

Traditional ethernet
frame   802.1Q frame

Figure 2.2 Traditional ethernet frame vs. 802.1Q frame.

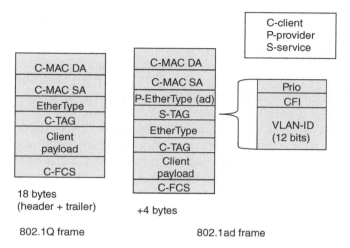

Figure 2.3 Provider bridges: IEEE 802.1ad (QinQ).

an Ethernet frame inside a service provider MAC header ("Mac-in-Mac"), the 802.1ah standard promises to provide support for over a million service instances, i.e. about 16 million; it also supports 802.1ad bridges; moreover, PBBs do not have to learn any customer's MAC address. This 802.1ah standard is also called ethernet virtual connections (EVC), Figure 2.4 shows the MAC in MAC data frame format.

Under the perspective of incremental upgrades of the Ethernet switching architecture to meet TE requirements, we can consider that the multiple

Figure 2.4 802.1ah frame.

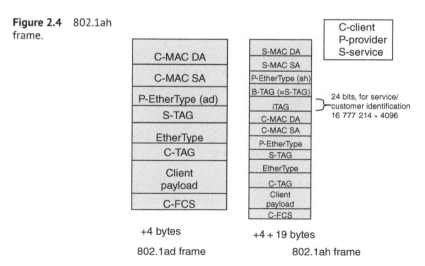

spanning tree protocol (MSTP) (Santos et al., 2011) has been the first attempt to actively perform TE in a legacy Ethernet switched network running STP or RSTP that suffers from unused links in normal situations. The multiplexing of multiple clients or VLANs into one among several spanning trees can also be optimized as presented in Santos et al. (2011). MSTP was originally inspired by the Cisco multiple instances spanning tree protocol (MISTP), to address the issue of unused links it introduces into the notion of multiple spanning tree (MST) region, a region that comprises several VLANs. Inside an MST there is a single internal spanning tree (IST). In practice, the IST corresponds to the regular spanning-tree obtained by running RSTP. In MSTP, the links blocked in an instance may be active in other instances. Hence, MSTP provides a better bandwidth efficiency. Other spanning-tree-based approaches were proposed in the literature as global open ethernet (GOE) (Iwata et al., 2004), or alternative multiple spanning tree protocol (AMSTP) (Ibanez et al., 2004).

Other protocols trying to solve bottleneck issues along the spanning tree(s) have been standardized, notably the Link Aggregation Group or the multi-chassis EtherChannel protocols (O'Reilly, 2011), allowing a switch to use multiple links as a single one with respect to the STP control-plane. Eventually, the real bottleneck in performing TE efficiently in an Ethernet switching context being the spanning tree bridging of Ethernet traffic, the STP control-plane has been removed from more recent carrier Ethernet solutions implementable in data-center networks (DCNs), namely the provider backbone bridges with traffic engineering (PBB-TE) (Wikipedia, 2009), where centralized control servers push MAC tables to backbone switches (in a similar philosophy OpenFlow (McKeown et al., 2008) does so too); the Layer 2 Label Switched Paths (L2LSP) (Papadimitriou et al., 2005) effort suggesting to use the VLAN fields as multiProtocol label switching (MPLS) label fields; the shortest path bridging (SPB) (Seaman, 2006), and transparent interconnection of a lot of links protocol (TRILL) (Touch and Perlman, 2009) protocols where the control-plane is distributed adapting a layer-3 link state routing protocol intermediate system to intermediate system (ISIS) to work with the Ethernet data-plane. Nodes in this context are no longer simple bridges since they perform a routing function. Hence, in TRILL, as well as in the following, we refer to them as router-bridges (referred to as RBridges, or RBs).

While differing in terms of scalability and deployability, the latter three solutions have proven to be viable ones and have been adopted by many vendors. Notably, these protocols enabled multipath routing of Ethernet frames, and hence opened the way to active load-balancing over multiple paths across virtual and physical switches.

Among these previous recent protocols designed in the last decade that include forms of multipath forwarding, also referred to as packet load-balancing techniques. They can act either at the data-link, network, or transport levels. At the data-link layer, the L2LSP was proposed to extend the capability of the generalized multiprotocol label switching (GMPLS) protocol based on Ethernet or any Layer 2 switching technology. More generally, this protocol uses the Ethernet frame capability for the data plane and the GMPLS for the control plane. Another protocol has been designed specifically for DCNs, the TRILL (Touch and Perlman, 2009). It allows a switch and even a virtualization server, acting as virtual bridge, to balance the load over multiple destination TRILL bridges for the same pair of nodes, it avoids a loop by using a time to live (TTL) field at layer 2. Similarly, the SPB IEEE 802.1aq also supports multipath forwarding (Seaman, 2006). However, no forms of congestion control are implemented with both SPB and TRILL, as the evolution of packet networks is such that this has been left to the transport layer.

At the network layer, equal cost multiPath (ECMP) (Thaler and Hopps, 2000) is adopted in the open shortest path first (OSPF) (Moy, 1998) and intermediate system to intermediate system (ISIS) protocols (Oran, 1990): it allows balancing the load over multiple next hops. ECMP can also be implemented in TRILL or SPB. However, this is typically performed in such a way that for a specific TCP flow, only one path is used in order to avoid packet disordering and buffer explosion at TCP endpoints.

Also at the network layer, we can cite locator/identifier separation protocol (LISP) (Farinacci et al., 2013), this protocol principally separates the IP function that leads the two functionalities into the identification and the localization. LISP separates these two functionalities; thus, LISP can be used as a control plane with IP data plane, and it can also be combined with border gateway protocol (BGP) (Rekhter and Li, 1995) as an example for forwarding packets inter autonomous system (AS).

Other protocols completely violate the TCP/IP architecture and propose an Ethernet over IP protocols as virtual extensible local area network (VXLAN) protocols, and the network virtualization using generic routing encapsulation (NVGRE). VXLAN(Sridhar et al., 2014), documented by the IETF in RFC 7348, is a network virtualization technology that attempts to ameliorate the scalability problems associated with large cloud computing deployments. It uses a VLAN to encapsulate layer 2 Ethernet frames within the layer 4 as user datagram protocol (UDP) packets. Hence, it allows for layer 2 adjacency across IP networks. VXLAN increases scalability up to 16 million logical networks and supports both multicast and unicast traffic using head-end replication (HER) to flood Broadcast, Unknown destination address, Multicast (BUM) traffic.

VXLAN is in reality an evolution of standardization efforts of the overlay encapsulation protocols. It was originally created by VMware, Arista Networks, and Cisco, and now is join by Broadcom, Citrix, Pica8, Cumulus Networks, Dell, Mellanox, OpenBSD, Red Hat, and Juniper Networks. Moreover, Open vSwitch supports VXLAN overlay networks.

NVGRE (Garg and Wang, 2015) is also a network virtualization technology that attempts to alleviate the scalability problems associated with large cloud computing deployments. NVGRE uses generic routing encapsulation (GRE) as the mechanism to virtualize IP addresses tunneling layer 2 packets over layer 3 networks. It has a 24-bit Virtual Subnet ID (VSID), which is stored in the GRE header of the new packet, that also allows scalability up to 16 million logical networks. The packet could have outer IP header as IPv6 and the inner IP header as IPv4. NVGRE-formatted packet has no inner TCP or UDP header. NVGRE is essentially endorsed by Microsoft, and it is also supported by Broadcom, Dell, Emulex, Intel, and Hewlett-Packard.

At the transport layer, there has been two major propositions. Stream control transmission protocol (SCTP) (Stewart, 2007) allows end hosts to use several paths concurrently, when devices are multihomed. The way it has been designed, however, makes SCTP weak against the de facto pervasive presence of middleboxes in the Internet such as firewalls, performance optimizers, load balancers at lower layers and interfaces. In many cases, SCTP connections cannot be opened or maintained. More recently, the multipath TCP (MPTCP) (Ford et al., 2011) has been designed with retrocompatibility and incremental deployability as the first design requirements so that using multiple paths simultaneously is made possible, falling back to standard TCP in case of middlebox blocking. Major attention has also be given to congestion control and fairness. An important requirement is that an MPTCP connection over a given link should not take more resources than legacy TCP connections running on the same link. However, as shown in Khalili et al. (2012), Medhi (2014), the congestion control algorithm is a key choice when fairness with respect to other connections needs to be guaranteed as it is a major concern of network operators. Finally, another proposed protocol Ethernet over TCP/IP is called stateless transport tunneling (STT) (Davie and Gross, 2013), a tunnel encapsulation that enables overlay networks to be built in virtualized networks. STT is particularly useful when some tunnel

Table 2.1 Summary of cloud network overlay protocols.

Cloud network overlay feature	SPB	TRILL	LISP	VXLAN	NVGRE	STT
Encapsulation	Ethernet over Ethernet	Ethernet over Ethernet	IP over IP	Ethernet over IP	Ethernet over IP	Ethernet over TCP/IP
Inter-datacenter link	Ethernet	Ethernet	IP	IP	IP	IP
Intra-datacenter link	Ethernet	Ethernet	IP	IP	IP	IP
User device integration	None	None	Yes	None	None	None
Virtual network segmentation	Yes	Limited	Yes	Yes	Yes	Yes
Firewall friendliness	Very high	Very high	High	High	Low	Very low
Incremental deployability	Low	High	Very high	High	Low	Low
Multipath and load balancing	Native	Native	Native	Partial	Partial	Partial
Multicast	Native	Native	Ongoing	Native	Partial	Partial

endpoints are in end-systems, as it utilizes the capabilities of the network interface card to improve performance.

Among the previous proposed protocols available at the Ethernet, we can cite SPB and TRILL; at the IP level, there is the LISP; and protocols on hybrid Ethernet-over-IP encapsulations include VXLAN, NVGRE, and STT. All these protocols natively support multipath forwarding and load balancing, at least to some extent. Table 2.1 briefly summarizes these six current cloud network overlay protocols. As the table suggests, they all natively support multipath forwarding and load balancing, at least to some extent. The most promising protocols, however, are also incrementally deployable, support virtual network segmentation, natively pass through IP networks, and easily cross firewalls.

2.2   Data Center Optimization Techniques

Based on the state-of-the-art, we can mainly consider three optimization domains: virtual network embedding, server consolidation, traffic engineering, and the TCP Proportional Fairness Model. Virtual network embedding (VNE) consists of mapping a set of virtual networks on a physical network, server consolidation consists of regrouping different virtualized nodes on the same hardware for saving energy and it is also sometimes referred to as energy efficiency (EE), the traffic engineering (TE), which is the application of technological and scientific principles as the measurement and modeling in order to control traffic in order to minimize maximum link utilization, for example another present challenge concerns the fairness in the offered throughput for the congestion control being. In this section, we will give an overview of these aforementioned techniques.

2.2.1   Virtual Network Embedding

The aim of VNE is to allow multiple virtual network topologies with widely varying characteristics to share the same physical hardware and the same physical networks in virtual data center (VDC), which consists of virtual switches and virtual machines (VMs) connected by virtual link.

A VDC can be seen as a logical instance of the Cloud Data Center or Virtualized Data Center, which is a Data Center that uses virtualization techniques. This Virtualized Data Center is a subset of the physical data center resources, called a substrate network (SN). VNE consists of embedding different virtual networks (VNs), the VNE primary element, on the substrate network. As shown in Figure 2.5.

VNE is often modeled as one Substrate Network (SN) as a graph with physical nodes and physical links, a set of Virtual Networks (VNs), where each VN is represented by a graph, vectors of the different considered resources depending on the authors, a function that associates each virtual node or links with its resource demands, a function that associates each physical node or link with its resource capacities, and finally functions that map the virtual node or link into the node or link.

We have observed different VNE approaches on the-state-of-the-art, static or dynamic approaches, centralized or distributed approaches. The static VNE approach does not consider the remapping possibility of some VNs to improve the embedding cost of the SN, while the dynamic approach allows a rem mapping based on the previous results and taking into consideration several metrics as SNs resources fragmentation. In fact, over time, some VNs can expire and release their resources from the SN,

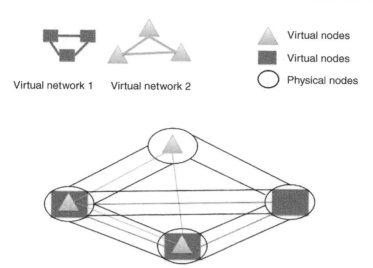

Figure 2.5 Virtual network embedding.

which causes fragmentation; trivially, this fragmentation will increase the future VN embedding requests. Hence, the dynamic approach allows better SN resources allocation; the authors of Fajjari et al. (2011b) have shown that most VN request rejections are due to bottlenecked links. A centralized approach means that one dedicated node (server) for example can be responsible for computing the VNE, while the distributed approach consists of collaborative works between different nodes to compute the VNE.

Regarding the metrics according to the performance goal, most of the study only consider one metric. Most of them consider the cost metric that minimize the mapping request cost. Others consider the Quality of Service metrics as the stress level, which is correlated to the amount of virtual nodes or links on the same physical nodes or links; the hop count or the path length; the utilization, which can concern the CPU, the RAM or the link; or finally the throughput. Some papers as Fajjari et al. (2011a) include two metrics in the objective function.

The VNE problem is known as an NP-hard problem. Indeed, authors adopt many algorithm strategies to solve it. The VNE problem can be seen as two subproblems, nodes mapping and link mapping. Hence, it can be solved by solving the two subproblems, some authors solve the two problems Independently also called the uncoordinated approach, or with a coordination with the two subproblems, where each solution affects the other.

Furthermore, many propositions have been proposed to solve the problem. Some proposed solutions have formulated the problem as a linear programming (LP) (Inführ and Raidl, 2011, Houidi et al., 2011), applicable only for small topologies and small instances of the problem. Many others study proposed heuristic solutions, Lischka and Karl (2009), Lu and Turner (2006), Houidi et al. (2008a), Houidi et al. (2008b), Marquezan et al. (2010), Yun and Yi (2011), or metaheuristic solutions, Fajjari et al. (2011a), solution that is a combination of an heuristic and LP solutions to solve the VNE problem.

We note that all the studies focusing on VDC are academic ones. These research fields did not receive any success in the industries. In fact, in reality, the ISP or Data Center providers manage their network globally using TE metrics, independently from the embedding nodes. However, embedding with the recent success of network function virtualization (NFV) architectures and with software defined networks (SDN) orchestration, it can be adopted by the industries for renting a virtual networks to customers. Recently, one study Guerzoni et al. (2014) proof this tendency, the authors proposed a novel approach by formulating the problem as mixed integer programming (MIP) problem that coordinated link and node mapping for software defined networks. It is worth noting that VNE strategies will be detailed in depth in Chapter 4.

2.2.2 Server Consolidation

VDC is a physical data center with deployed resource virtualization techniques, which means that VMs are installed on servers. Server consolidation, often modeled as VM placement, aims to reduce the number of used servers by regrouping the VMs on the same server whenever possible.

This VM placement is an important approach for improving energy efficiency and resource utilization in cloud infrastructures. Most of the works have modeled VM consolidation problem, and considered it as an NP-hard optimization problem. Hence, many works have proposed a heuristic to solve it. Some of them proposed extensions of simple greedy algorithms such as First Fit Decreasing (Wood et al., 2009), Least full first, and Most Full First (Lee et al., 2011), Best Fit (Mishra and Sahoo, 2011).

In Meng et al. (2010), the authors proposed a VM placement solution considering network resource consumption. They designed a two-tier approximation algorithm that efficiently solves the VM placement problem. They assumed that a VM container could be divided into CPU-memory slots, where each slot could be allocated to any VM. They considered the number of VMs as equal to the number of slots; if the number of slots was higher than the number of VMs, they added dummy VMs (with no traffic), and did not affect the algorithm. Due to a communication cost between slots, defined as the number of forwarded frames among them, the objective was set as the minimization of the average forwarding latency. They also assumed static single-path routing and focused on two traffic models. A dense one where each VM sent traffic to every VMs at an equal and constant rate, and a sparse infrastructure as a service (IaaS)-like one with isolated clusters so that only VMs in the same IaaS could communicate.

Some other works proposed a genetic algorithm to solve the problem. In Mi et al. (2010), the authors proposed a genetic algorithm-based approach, namely genetic algorithm based reconfiguration (GABA), the authors proposed an online self-reconfiguration approach for reallocating VMs in large-scale data centers, and just considered the CPU resources. In Xu et al. (2010), the authors also propose a genetic algorithm for resolving the VM placement problem. They formulate the problem as a multiobjective optimization minimizing the total resource wastage, the power consumption, and thermal dissipation costs, and they proposed genetic algorithm to incrementally search the better solution combining possibly conflicting objectives.

Many works have formulated the problem as a bin-packing problem. In Wang et al. (2011), the authors consolidated VM placement considering a nondeterministic estimation of bandwidth demands. The bandwidth demand of VMs was set to follow normal distributions as the authors assumed that server consolidation usually occurs at weekly or monthly timescale. They formulated the consolidation in a Stochastic Bin Packing problem and introduced an online heuristic approach to resolve it. In Li et al. (2009), the authors also formulated the problem as a bin-packing problem considering only server resources constraints, and they proposed an optimized First-Fit-Decreasing, where instead of placing directly the new arrived workload into the first node that can accommodate, the heuristic tries to reorganize the workloads smaller than the newcomer. The heuristic also reorganizes the workloads when workload departs.

In Rabbani et al. (2013), the authors revisited the virtual embedding problem by distinguishing between server and bridge nodes with respect to the common formulation. They proposed an iterative three-step heuristic: during the first step, an arbitrary VM mapping was done; the second step mapped virtual bridges to bridge nodes, and the third one mapped virtual links accordingly. If one of these steps failed, the heuristic would come back to the previous one until a solution was found. The quality of the solution seemed dependent on the first step, the other steps just minimized the impact of the previous step. Furthermore, there may have been a scaling problem due to the uncontrollable backtracking. More generally, virtual embedding approaches in the literature often discarded specificities of the network control-plane such as the routing protocol and TE capabilities.

In Biran et al. (2012), the authors considered network constraints in addition to CPU and memory constraints in the VM placement problem. They defined a network-aware VM placement optimization approach to allocate VM placement while satisfying predicted traffic patterns and reducing the worst-case cut load ratio in order to support time-varying traffic. Interested by network cuts, they partitioned the set of hosts into nonempty connected subsets, which are bottlenecks for the traffic demand between VMs placed in different sides of the cut.

In Jiang et al. (2012), the authors optimized jobs placements where each job required a number of VMs; the objective function minimized the network and the node costs. The authors did not handle the link capacity constraints and did not consider multipath forwarding capabilities instead of multipath routing with one single egress path.

In Jin et al. (2013), the authors minimized the power energy consumption of activated servers, bridges, and links to maximize the global energy saving. The authors converted the VM placement problems into a routing problem, and so they addressed the network and server optimization problem as a single one. So there was no trade-off between the network-side and server-side optimization objective.

Only few recent studies have proposed a metaheuristic to solve server consolidation problem. In Gao et al. (2013), the authors proposed Ant Colony Optimization a metaheuristic to minimize two objectives: the resource wastage and the power consumption. Also, in Ferdaus et al. (2014), they solved the problem by using Ant Colony Optimization and only considered CPU and memory constraints.

Most of the studies on the state-of-the-art focused on a single criterion. Some of these studies ignored link capacity constraints, Wood et al. (2009), Lee et al. (2011), Lee et al. (2011), Mishra and Sahoo (2011), others excluded dynamic routing as in Meng et al. (2010), or just considered the traffic volume to reduce the number of containers as in Wang et al. (2011), or just the network resources as in Meng et al. (2010) and Biran et al. (2012), only Jin et al. (2013) considered multipath forwarding capabilities.

Some of these studies excluded dynamic routing; for example in Meng et al. (2010) the authors propose a VM placement solution considering network resource consumption, wherein the objective is set as the minimization of the average forwarding latency. Other studies take only the network resources into account, as in Meng et al. (2010), Biran et al. (2012), or just consider the traffic volume to reduce the number of containers as in Wang et al. (2011), where authors propose a VM placement considering a nondeterministic estimation of bandwidth demands, formulating the problem as a Stochastic Bin Packing problem, and introducing a new heuristic approach to resolve it. In Rabbani et al. (2013), where the authors revisited the virtual embedding problem by distinguishing between server and switching nodes; they do not handle the link capacity constraints, and they also do not consider multipath forwarding (load balancing), but a multipath routing with single egress path.

Jin et al. (2013) was the first work where the authors minimized the energy consumption of active servers, bridges, and links to maximize the global EE. The authors converted the VM placement problem into a routing problem, so as to address the joint network and server optimization problem (where there is no tradeoff between network and server objectives).

Commonly, because of the relatively recent employment of virtual bridging for transiting traffic at the server level, virtual bridging capabilities for external traffic forwarding were ignored. The first studies that have been designed and proposed a metaheuristic considering the tread-off between network and server consolidation objectives under traffic engineering and energy efficiency constraints and considering multipath forwarding was Belabed et al. (2014a), Belabed et al. (2014b), where the first one focus on the impact of Virtual Bridging on server consolidation and the second one on the impact of Ethernet Multipath Routing. Finally, the authors have extented their work and investigated the impact of virtualization servers and Multipath Routing in DCN in Belabed et al. (2015).

2.2.3 Traffic Engineering

Traffic Engineering (TE) is not a new science; Internet Traffic Engineering consists of applying traffic engineering theory to telecommunications. As defined by Awduche et al. (2002)

> "Internet traffic engineering is defined as that aspect of Internet network engineering dealing with the issue of performance evaluation and performance optimization of operational IP networks. Traffic Engineering encompasses the application of technology and scientific principles to the measurement, characterization, modeling, and control of Internet traffic…".

Internet Service Providers and Data Center providers aim, by using TE techniques, for better traffic control since TE allows to route a particular traffic in specific maker base on source–destination routing while IP routing is based only on destination, avoiding congestion, reducing capital expenditure (CAPEX) cost, and optimizing the global revenue.

In this subsection, we are going to overview some known traffic engineering methods, Traffic engineering in link-state routing, MPLS and traffic engineering, Inter-AS TE, and BGP-TE methods.

2.2.3.1 Link-State Traffic Engineering

Link-state protocol is based on each node (router) of the graph having to own a global view of the topology, hence each node can compute the shortest path by using a specific algorithm as Dijkstra, to fill its own routing table. The routing information as reachability, cost, or link state are periodically exchanged between routers thus to obtain the global topology view. Indeed, if in case of a failure, a link-state is broadcasted, each router updates its topological data base and along the way update its routing table.

Link-state TE is based on OSPF and ISIS the two popular interior gateway protocol (IGP) protocols. Using TE in link-state routing consists of handling the IGP administrative weights that will influence the result of the algorithm of ISIS or OSPF mechanisms. Weights can be used can be assigned to avoid congestion, to maximize or minimize the network utilization, to differentiate high-speed links from the slow ones.

Traffic Engineering in Link-State IP Networks can be summarized as a link weights optimization in order to meet some objectives. TE in Link-State IP Networks is often modeled as a mathematical problem that aims to minimize the Maximum Link Utilization taking into account network constraints.

Furthermore, load sharing can also be used as TE mechanism. Indeed, ECMP is supported by IGP protocols, in order to avoid packet sorting by the destination, load sharing is usually implemented per flow (not per packet).

Obviously, Link-state TE applies for intradomain, the time convergence based on IGP, from link failure, requires is about five seconds, and it is done one five steps, failure detection (Hellos, SDH alarms), Link-State broadcast, topological data base update, shortest path recomputation, and Forwarding Information Base (FIB) update.

2.2.3.2 MPLS Traffic Engineering

MPLS was created in order to manage IP packets as a Data Plan. In fact, we can map MPLS technology on Layer 2 technology that divides the control plan from the data plan, where the role of the data plan consists of basic functions such as filtering, forwarding, and flooding for the frames. MPLS separates the data paths from the control paths at Layer 3. This is why MPLS is often rolled as a 2.5 technology.

At MPLS, the data paths are identified using labels, we can see the analogy with the layer 2 VLAN tag, we note that originally MPLS payload focused on IP, after that a GMPLS (Mannie, 2004) standard came to regroup the different MPLS works as MPLS over synchronous optical netwoort/synchronous digital hierarchy [time-division multiplexing] (SONET/SDH [TDM]) or Optical (Lambdas).

MPLS is based on controlling data paths called Label Switched Path (LSP), there are mainly three LSP types: the prefix based: the LSPs is created based on route advertisement controlled by routing or signaling; the tunnel based: LSPs created between specific MPLS end-points.

MPLS-TE used the LSP tunneling technique and consists of an explicit route selection. It basically consists of two protocols: OSPF-TE or ISIS-TE (Chen et al., 2008), and Resource ReSerVation Protocol TE (RSVP-TE) (Awduche et al., 2001), as shown on Figure 2.6. OSPF-TE or ISIS-TE consists of an extension of the previous protocols by giving more information about the topology, which can be exchanged as the available bandwidth, the information can be exchanged using opaque link-state

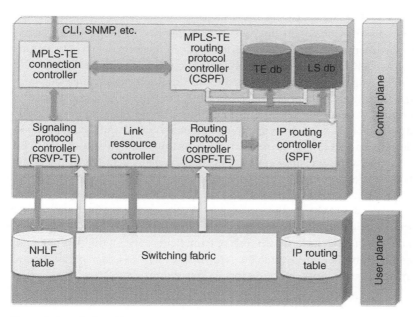

Figure 2.6 MPLS traffic engineering.

advertisement (LSA) carrying type-length-value elements. OSPF-TE (Katz et al., 2003) is used in GMPLS networks to get the global view of the topology, and it is completely out of band of the data plane network. Hence, it can also be used on non-IP networks, such as optical networks. RSVP-TE is also an extension of RSVP taking into consideration more network information constraint parameters such as available bandwidth and explicit hops. The RSVP-TE specification generalizes the concept of RSVP to flow, in fact, originally the RSVP concept was created to be used by hosts to reserve the required resources for micro flows based on forwarding equivalence class (FEC), while RSVP-TE allows RSVP session to consist of an arbitrary aggregation of traffic based on local policies between the ingress node of an LSP-tunnel and the egress node of the tunnel.

Regarding load balancing, several LSPs can be opened for one FEC, thus each path can carry some traffic percentage. Furthermore, MPLS-TE (Kompella et al., 2005) recovery technique is fast and efficient, it is based on LSP protection that use preallocated back-up resources or LSP restoration that does not preallocate resources but can precompute the route, and use presignaled LSP back-up, by reserving resource without activating LSP.

2.2.3.3 TCP Proportional Fairness Model

Since the first steps of the Internet, it was recognized that unrestricted access to the Internet resulted in poor performance in the form of low network utilization and high packet loss rates (Jacobson, 1988). These phenomena were called congestion collapses and led to the development of congestion control algorithms appropriate to the Internet. The basic dominating idea was to detect the congestion in the network through packet losses and integrate this information in the way throughput is offered to greedy applications; upon detecting a packet loss, the source reduces its transmission rate, or otherwise it increases the transmission rate. Eventually, many versions of the transport control protocol (TCP) (Postel, 1981) use the lack of acknowledgment of packets as an alert of packet loss.

In a network with multiple competing TCP sessions sharing links, several studies Srikant (2004), Kunniyur and Srikant (2003), Wei et al. (2006), and Low et al. (2002) have shown that TCP implicitly solves a utility problem in equilibrium. This utility problem is formally described as a maximization of an aggregate utility subject to capacity constraints:

$$\max_{X \geq 0} \sum_{j \in \mathcal{J}} U(X_j) \tag{2.1}$$

subject to

$$\sum_{j \in \mathcal{J}} \delta_{je} X_j \le c_e, \quad e = 1, 2, \dots, E \tag{2.2}$$

The above model maximizes the utility function $U(X_j)$ of each session $j \in \mathcal{J}$, where X_j denotes the rate of session j, while δ_{je} is the indicator that takes the value 1 if session j uses link e, 0, otherwise.

Multiple approaches in order to address the fairness between costumers exist. In Pióro and Medhi (2004), two approaches are described, the *max–min fairness* (MMF) approach, which maximizes the minimal assignment of capacity to demands. In fact the solution can be resumed in iterative steps; first, MMF assigns the same minimal volume to all demands, until some free capacity is still present, each minimal assignment is increased whenever possible, and so on until the maximization does not improve any longer the situation.

To illustrate MMF, the authors consider a network, where V is the set of nodes, E is the set of links and D is the set of demands between nodes. Each node is identified by v_i such as $v_i \in V$ and $i = 0, 1, \dots, |V|$ and each link is identified by e_j such as $e_j \in E$ and $j = 0, 1, \dots, |E|$ and finally each demand is identified by d_k such as $d_k \in D$ and $k = 0, 1, \dots, |D|$. Each demand is identified by an origin and a destination between two nodes. A route is specified by a sequence of links. Let x_d be the rate at which source v is allowed to transmit data. Each link e in the network has a capacity c. Given the capacity constraints on the links, the source allocation problem is to assign a rate x_d to the users in a fair manner.

To illustrate the difficulties in defining a fair allocation, consider the example in Figure 2.7. Suppose a network with three links, two nodes, and three demands. The network consists of two links e_1 and e_2; and three nodes v_1, v_2, and v_3; and three demands d_1, d_2, and d_3. d_1 and d_2 cover only one link, e_1 and e_2, respectively. The third demand d_3 covers both links, e_1 and e_2. Suppose that the capacity of link e_1 is $c_1 = 2$, of link e_2 is

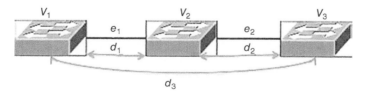

Figure 2.7 A reference network example for MMF and PF allocations. Source: Pióro and Medhi (2004).

$c_2 = 1$. A resource allocation that satisfies the link capacity constraints is $x_2 = x_3 = 0.5$ and $x_1 = 1.5$.

Suppose now that you attempt to divide the capacity of each link among the demands using the link. Then, on link e_1, demands d_1 and d_3 would get a rate of 1 each and on link e_2, demand d_2 and d_3 would get a rate of 0.5 each. However, demand d_3 can only transmit at rate 0.5 because it covers both links e_1 and e_2. Thus, there is still one more unit of capacity remaining to be allocated on link e_1. This remaining capacity is now allocated to the only other demand using that link, which is demand d_1, thus, given $x_1 = 1.5$, $x_2 = 0.5$, and $x_3 = 0.5$.

Then, supposing that $c_1 = c_2 = 1.5$, the resource allocation that satisfies the link capacity constraints is $x_1 = x_2 = x_3 = 0.75$. Clearly, this is not optimal with respect to the throughput. In fact, the resulting throughput is equal to $x_1 + x_2 + x_3 = 2.25$.

To the best of our knowledge, only Belabed et al. (2014c) tries to address the impact of topology design, capacity planning, and multipath forwarding in traffic fairness in DCN fabrics.

In the alternative *proportional fairness* (PF) Mazumdar et al. (1991), Kelly et al. (1998) allocation that is applicable to TCP, utility $U(x_j)$ is set to $\omega_j \log x_j$, where ω_j is the weight of the session j. Hence, the resource allocation corresponding to this utility function is commonly referred to as *weighted proportionally fair*, or, if all ω_d are equal to one, as *proportionally fair*. Thus, (2.1) for PF becomes

$$\max_{X \geq 0} \sum_{j \in J} \omega_j \log X_j \tag{2.3}$$

The *log* function avoids the assignment of zero or too low volumes to demands since $\log x \to -\infty$ as $x \to 0$. Furthermore, it makes it not profitable to assign much volume to any demand. We note that the PF objective function is nonlinear with respect to the MMF one. Taking the example in Figure 2.7, the corresponding explicit resource allocation is given by

$$max \{ \log x_1 + \log x_2 + \log x_3 \} \tag{2.4}$$

subject to:

$$x_1 + x_3 \leq 2 \tag{2.5}$$

$$x_2 + x_3 \leq 1 \tag{2.6}$$

$$x_1, x_2, x_3 \in R^+ \tag{2.7}$$

The corresponding weighted proportionally fair resource allocation is $c_1 = c_2 = 1.5$, which favor shortest flows. The resource allocation that satisfies the link capacity constraints is $x_1 = x_2 = 1$ and $x_3 = 0.5$. From a fairness perspective, the PF solution is less fair than the MMF solution. However, since it favors shorter flows, the PF allocation is more efficient in terms of global throughput, in this case it is equal to $x_1 + x_2 + x_3 = 2.5$.

2.3 Conclusion

In the context of DCN optimization, virtual bridging becomes very useful for the management of VMs co-located in the same virtualization server, by offloading inter-VM traffic from access and aggregation switches, at the expense of an additional computing load on the virtualization server. Moreover, with the emergence of flat DCN topologies, such as Fat-Tree Al-Fares et al. (2008), DCell Guo et al. (2008), and BCube Guo et al. (2009), multipath forwarding can become very useful to fully utilize the available paths and capacity, and therefore offer higher throughput and resiliency to servers. The ability to synchronize VM copies and migrate across virtualization servers further adds elasticity to the cloud fabric, by allowing on one hand fault-restoration, and massive distributed resource consolidation on the other.

However, for instance, migrating a VM catalyst of significant traffic at ("VM containers") servers whose access link is close to saturation is not a wise decision. VM containers that are topologically attractive could therefore be favored when deciding where to host and migrate VMs. The most commercial DCN consolidation tools, typically are aware of CPU, memory, storage, and energy constraints of VM containers, are not aware of link states since the legacy hypothesis is to consider unlimited link capacity. With the emergence of network virtualization, related storage synchronization tools and pervasive virtual bridging, the hypothesis that DCN links have infinite capacity is today becoming inappropriate, especially for DCs facing capital expenditure limitations. Performing VM consolidation that is aware of both container and link states is, however, known to be NP-hard (Zhang et al., 2010). The complexity does naturally increase when considering multipath and virtual bridging capabilities.

In this context, the recent introduction and large deployment of virtual bridging in most hypervisor solutions (e.g. XeN, KVM, VM Ware NSX), is introducing novel constraints as it becomes interesting to assign to a same container or nearby containers VMs exchanging large traffic amounts. The impact of virtual bridging on DCN consolidations can be sensible, depending on topology and forwarding situations.

Virtual machine placement typically chases Traffic Engineering (TE) Jiang et al. (2012), Fang et al. (2013) and Energy Efficiency (EE) Van et al. (2010), Hermenier et al. (2009) objectives, i.e. it tries to minimize maximum link utilization when balancing the traffic load on DCN links, and maximize server utilization to turn off or hibernate some servers to save energy. The interrelationship between these three recent trends (virtual bridging,[1] multipath forwarding[2] and VM placement) is difficult objective to achieve (Belabed et al., 2015).

Another present challenge in Data Center networks is to better understand the impact of novel flattened and modular DCN architectures on congestion control protocols, and vice versa. One of the major concerns in congestion control being the fairness in the offered throughput, the impact of the additional path diversity and forwarding features, brought by the novel DCN architectures and protocols, on the throughput of individual endpoints (servers) and aggregation points (edge switches) is unclear.

With the growth in customer voluminous, service differentiated and elastic demands, avoiding bottlenecks is a critical point in Data Center network architectures. With the de facto dominating trend of deploying services using virtualization servers, a nonnegligible ratio of the traffic is horizontal traffic between virtualization servers, in support of VM migration and storage synchronization. The amount of intra-DC horizontal traffic can overcome the access vertical traffic volume (Benson et al., 2010). This has eventually favored the emergence of novel DCN architectures that expose additional horizontal capacity between server racks and clusters of racks such as fat-tree (Al-Fares et al., 2008), and BCube Guo et al. (2009).

An open question is: how best is the traffic allocation of the competing elastic demand flows for horizontal traffic between edge servers in Data Center fabrics, and how is this allocation impacted with the increase in capacity? Only few studies try to understand this impact in equilibrium Belabed et al. (2014c) and Hu et al. (2015).

The purpose of this chapter is to provide a deeper understanding of the Data Center network fabrics. We investigate the impact of the novel features in DCN optimization, providing comprehensive definitions. We show in particular how virtual bridging and multipath forwarding impact common DCN optimization goals "Traffic Engineering and Energy Efficiency". After determining the highly virtual bridging and multipath forwarding impact on

---

1 Virtual bridging: is a bridge at the hypervisor software of the container. Hence, the traffic between VMs at the container, incoming and outgoing can be managed at the container instead of the physical bridges.
2 Multipath forwarding: allowed the use of more than one path at the same time for carrying the traffic.

DCN protocols and on optimization modeling "Virtual network embedding and server consolidation", our interest moves into investigating the fairness of the Data Center network fabrics and how is it impacted by the multipath.

Bibliography

M. Al-Fares, A. Loukissas, and A. Vahdat. A scalable, commodity data center network architecture. 2008.

D. Awduche, L. Berger, D. Gan, T. Li, V. Srinivasan, and G. Swallow. RSVP-TE: extensions to RSVP for LSP tunnels. *RFC 3209*, 2001.

D. Awduche, A. Chiu, A. Elwalid, I. Widjaja, and X.P. Xiao. Overview and principles of internet traffic engineering. *RFC 3272*, 2002.

D. Belabed, S. Secci, G. Pujolle, and D. Medhi. Impact of virtual bridging on virtual machine placement in data center networking. 2014a.

D. Belabed, S. Secci, G. Pujolle, and D. Medhi. Impact of ethernet multipath routing on data center network consolidations. 2014b.

D. Belabed, S. Secci, G. Pujolle, and D. Medhi. On traffic fairness in data center fabrics. 2014c.

D. Belabed, S. Secci, G. Pujolle, and D. Medhi. Striking a balance between traffic engineering and energy efficiency in virtual machine placement. *IEEE Transactions on Network and Service Management*, 2015.

T. Benson, A. Akella, and D.A. Maltz. Network traffic characteristics of data centers in the wild. 2010.

O. Biran, A. Corradi, M. Fanelli, L. Foschini, A. Nus, D. Raz, and E. Silvera. A stable network-aware VM placement for cloud systems. 2012.

S. Bradner. The internet engineering task force. *IETF*, 1999.

Bridges. Provider Backbone Bridges. IEEE standard 802.1ah. 2008.

M. Chen, R. Zhang, and X. Duan. ISIS extensions in support of inter-autonomous system (AS) MPLS and GMPLS traffic engineering. *RFC 5316*, 2008.

B. Davie and J. Gross. A stateless transport tunneling protocol for network virtualization (STT). *draft-davie-stt-04*, 2013.

EK Niclas. 802.1 P,Q-QoS on the MAC level. *IEEE*, 1999.

I. Fajjari, N. Aitsaadi, G. Pujolle, and H. Zimmermann. VNE-AC: virtual network embedding algorithm based on ant colony metaheuristic. 2011a.

I. Fajjari, N. Aitsaadi, G. Pujolle, and H. Zimmermann. VNR algorithm: a greedy approach for virtual networks reconfigurations. 2011b.

W. Fang, X. Liang, S. Li, L. Chiaraviglio, and N. Xiong. VMPlanner: optimizing virtual machine placement and traffic flow routing to reduce network power costs in cloud data centers. *Computer Networks*, 2013. 57. 179–196.

D. Farinacci, D. Lewis, D. Meyer, and V. Fuller. The locator/ID separation protocol (LISP). 2013.

M.H. Ferdaus, M. Murshed, R.N. Calheiros, and R. Buyya. Virtual machine consolidation in cloud data centers using ACO metaheuristic. 2014.

A. Ford, C. Raiciu, M. Handley, S. Barre, and J. Iyengar. Architectural guidelines for multipath TCP development. *IETF RFC6182*, 2011.

Y. Gao, H. Guan, Z. Qi, Y. Hou, and L. Liu. A multi-objective ant colony system algorithm for virtual machine placement in cloud computing. *Journal of Computer and System Sciences*, 2013. 79. 1230–1242,

P. Garg, Y. Wang, et al. IETF RFC 7637: NVGRE: Network virtualization using generic routing encapsulation. 2015.

R. Guerzoni, R. Trivisonno, I. Vaishnavi, Z. Despotovic, A. Hecker, S. Beker, and D. Soldani. A novel approach to virtual networks embedding for SDN management and orchestration. 2014.

C. Guo, H. Wu, K. Tan, L. Shi, et al. DCell: a scalable and fault-tolerant network structure for data centers. 38 (4): 75–86, 2008.

C. Guo, G. Lu, D. Li, H. Wu, X. Zhang, Y. Shi, C. Tian, Y. Zhang, and S. Lu. BCube: a high performance, server-centric network architecture for modular data centers. *ACM SIGCOMM Computer Communication Review*, 2009. 39. 63–74,

F. Hermenier, X. Lorca, J.-M. Menaud, G. Muller, and J. Lawall. Entropy: a consolidation manager for clusters. 2009.

I. Houidi, W. Louati, and D. Zeghlache. A distributed virtual network mapping algorithm. 2008a.

I. Houidi, W. Louati, and D. Zeghlache. A distributed and autonomic virtual network mapping framework. 2008b.

I. Houidi, W. Louati, W.B. Ameur, and D. Zeghlache. Virtual network provisioning across multiple substrate networks. *Computer Networks*, 2011. 55. 1011–1023.

Z. Hu, Y. Qiao, and J. Luo. ATME: accurate traffic matrix estimation in both public and private datacenter networks. *IEEE Transactions on Cloud Computing*, 2015. 6. 60–73.

G. Ibanez, A. Garcia, and A. Azcorra. Alternative multiple spanning tree protocol (AMSTP) for optical ethernet backbones. 2004.

IEEE Std 802.1ad-2005. IEEE standard for local and MANs—virtual bridged local area networks—amendment 4: provider bridges. *(Amendment to IEEE Std 802.1Q-2005)*, 2006.

J. Inführ and G.R. Raidl. Introducing the virtual network mapping problem with delay, routing and location constraints. 2011.

A. Iwata, Y. Hidaka, M. Umayabashi, N. Enomoto, and A. Arutaki. Global open ethernet (GOE) system and its performance evaluation. *IEEE Journal on Selected Areas in Communications*, 2004. 22. 1432–1442.

V. Jacobson. Congestion avoidance and control. 1988.

J.W. Jiang, T. Lan, S. Ha, M. Chen, and M. Chiang. Joint VM placement and routing for data center traffic engineering. 2012.

H. Jin, T. Cheocherngngarn, D. Levy, A. Smith, D. Pan, J. Liu, and N. Pissinou. Joint host-network optimization for energy-efficient data center networking. *IEEE IPDPS*, 2013.

D. Katz, K. Kompella, and D. Yeung. Traffic engineering (TE) extensions to OSPF version 2. *RFC 3630*, 2003.

F.P. Kelly, A.K. Maulloo, and D.K.H. Tan. Rate control for communication networks: shadow prices, proportional fairness and stability. *Journal of the Operational Research society*, 1998. 49. 237–252.

R. Khalili, N. Gast, M. Popovic, U. Upadhyay, and J.-Y.L. Boudec. MPTCP is not pareto-optimal: performance issues and a possible solution. 2012.

K. Kompella, L. Berger, and Y. Rekhter. Link bundling in MPLS traffic engineering (TE). 2005.

S. Kunniyur and R. Srikant. End-to-end congestion control schemes: utility functions, random losses and ECN marks. *IEEE/ACM Transactions on Networking*, 2003. 11. 689–702

S. Lee, R. Panigrahy, V. Prabhakaran, V. Ramasubramanian, K. Talwar, L. Uyeda, and U. Wieder. Validating heuristics for virtual machines consolidation. *Microsoft Research, MSR-TR-2011-9*, 2011.

B. Li, J. Li, J. Huai, T. Wo, Q. Li, and L. Zhong. EnaCloud: an energy-saving application live placement approach for cloud computing environments. 2009.

J. Lischka and H. Karl. A virtual network mapping algorithm based on subgraph isomorphism detection. 2009.

S.H. Low, L.L. Peterson, and L. Wang. Understanding TCP Vegas: a duality model. *Journal of the ACM*, 2002. 49. 207–235

J. Lu and J. Turner. Efficient mapping of virtual networks onto a shared substrate. 2006.

E. Mannie. Generalized multi-protocol label switching (GMPLS) architecture. IETF RFC 3945. 2004.

C.C. Marquezan, L.Z. Granville, G. Nunzi, and M. Brunner. Distributed autonomic resource management for network virtualization. 2010.

R. Mazumdar, L.G. Mason, and C. Douligeris. Fairness in network optimal flow control: optimality of product forms. *IEEE Transactions on Communications*, 1991. 39. 775–782

N. McKeown, T. Anderson, H. Balakrishnan, et al. OpenFlow: enabling innovation in campus networks. *ACM SIGCOMM Computer Communication Review*, 2008. 38. 69–74

D. Medhi. Applications with multiple parallel flows: assessing their unfair advantage with proportional fair sharing TCP. 2014.

X. Meng, V. Pappas, and L. Zhang. Improving the scalability of data center networks with traffic-aware virtual machine placement. 2010.

H. Mi, H. Wang, G. Yin, Y. Zhou, D. Shi, and L. Yuan. Online self-reconfiguration with performance guarantee for energy-efficient large-scale cloud computing data centers. 2010.

M. Mishra and A. Sahoo. On theory of VM placement: anomalies in existing methodologies and their mitigation using a novel vector based approach. 2011.

J. Moy. Open shortest path first (OSPF) version 2. *IETF: The Internet Engineering Taskforce RFC 2328*, 1998.

D. Oran. OSI IS-IS intra-domain routing protocol. 1990.

Gary A. Donahue (2011). Multi-chassis ether channel (MEC). In: *Network Warrior* O'Reilly Media, Inc.

D. Papadimitriou, E. Dotaro, and M. Vigoureux. Ethernet layer 2 label switched paths. 2005.

R. Perlman. An algorithm for distributed computation of a spanningtree in an extended LAN. *ACM SIGCOMM Computer Communication Review*, 1985. 15. 44–53.

M. Pióro and D. Medhi. Routing, flow, and capacity design in communication and computer networks. 2004.

J. Postel. Transmission control protocol. 1981.

M.G. Rabbani, R.P. Esteves, M. Podlesny, G. Simon, L.Z. Granville, and R. Boutaba. On tackling virtual data center embedding problem. 2013.

Y. Rekhter and T. Li. A border gateway protocol 4 (BGP-4). 1995.

D. Santos, A. de Sousa, F. Alvelos, M. Dzida, and M. Pióro. Optimization of link load balancing in multiple spanning tree routing networks. *Telecommunication Systems*, 2011.48. 109–124.

M. Seaman. IEEE 802.1aq shortest path bridging. *IEEE Draft Standard*, 2006.

T. Sridhar, L. Kreeger, D. Dutt, C. Wright, M. Bursell, M. Mahalingam, P. Agarwal, and K. Duda. Virtual extensible local area network (VXLAN): a framework for overlaying virtualized layer 2 networks over layer 3 networks. 2014.

R. Srikant. The mathematics of internet congestion control. 2004.

R. Stewart. Stream control transmission protocol. 2007.

D. Thaler and C. Hopps. Multipath issues in unicast and multicast next-hop selection. *RFC 2991*, 2000.

P. Thaler, N. Finn, D. Fedyk, G. Parsons, and E. Gray. IEEE 802.1q. *IEEE*, 2013.

J. Touch and R. Perlman. Transparent interconnection of lots of links (TRILL): problem and applicability statement. *RFC 5556*, 2009.

H.N. Van, F.D. Tran, and J.-M. Menaud. Performance and power management for cloud infrastructures. 2010.

W. Vojdak. Rapid spanning tree protocol: a new solution from an old technology. *Compact PCI Systems*, 2008.

M. Wang, X. Meng, and L. Zhang. Consolidating virtual machines with dynamic bandwidth demand in data centers. 2011.

D.X. Wei, Cheng Jin, S.H. Low, and S. Hegde. FAST TCP: motivation, architecture, algorithms, performance. *IEEE/ACM Transactions on Networking*, 2006. 14. 1246–1259,

IEEE Standard for Local and metropolitan area networks – Virtual Bridged Local Area Networks Amendment 10: Provider Backbone Bridge Traffic Engineering. *IEEE Std 802.1Qay-2009 (Amendment to IEEE Std 802.1Q-2005)*, 2009. 1–142, doi: 10.1109/IEEESTD.2009.5198465.

T. Wood, P. Shenoy, A. Venkataramani, and M. Yousif. Sandpiper: black-box and gray-box resource management for virtual machines. *Computer Networks*, 2009.53. 2923–2938.

J. Xu, F., and J. AB. Multi-objective virtual machine placement in virtualized data center environments. 2010.

D. Yun and Y. Yi. Virtual network embedding in wireless multihop networks. 2011.

Q. Zhang, L. Cheng, and R. Boutaba. Cloud computing: state-of-the-art and research challenges. *Journal of Internet Services and Applications*, 2010. 1. 7–18.

3

Resource Management in Hybrid (Wired/Wireless) Data Center Networks

Boutheina Dab[1,2], Ilhem Fajjari[3], and Nadjib Aitsaadi[4]

[1] *VMware, Hauts-de-Seine, La Defense, France*
[2] *LiSSi Lab, UPEC, Val de Marne, Vitry sur Seine, France*
[3] *Orange Labs, Orange, Hauts-de-Seine, Chatillon, France*
[4] *Universités Paris-Saclay, UVSQ, DAVID, F-78035, Versailles, France*

Abstract

Routing and resource allocation are key challenges in hybrid data center networks. Ensuring an efficient management of wireless and wired infrastructure in the hybrid (wired/wireless) data center networks (HDCN), for both one-hop and multi-hop communications, is primordial to guarantee a high performance network. In one-hop inter-rack communications, the sending and receiving racks are placed in the same wireless transmission range, the objective is to find efficient algorithms for wireless channel allocation in HDCN while minimizing the congestion level. Several recent research approaches have explored the feasibility of deploying wireless links in HDCN based on practical testbeds, but only few studies have been conducted to perform wireless channel allocation.

On the other hand, the multi-hop inter-rack communications require efficient mechanisms to jointly route and allocate channels for the communication flows, while enhancing network performance. The objective is to compute for each flow, the hybrid (i.e. wireless and/or wired) routing path. In this regard, the joint routing and wireless channel allocation problem in HDCN can be addressed either in an online or a batch way. In the online mode, inter-rack communication flows are sequentially processed in order to find the hybrid routing path for each single flow request. Few research works have been proposed to deal with this issue. However, even if the online approaches guarantee an optimized hybrid routing path for each single flow request, they fail to ensure an optimized use of the wireless and wired

Management of Data Center Networks, First Edition. Edited by Nadjib Aitsaadi.
© 2021 The Institute of Electrical and Electronics Engineers, Inc.
Published 2021 by John Wiley & Sons, Inc.

resources in the HDCN. Indeed, the arrival order closely impacts the HDCN performance. Therefore, a few recent researches have investigated the joint batch routing and channel (JBRC) assignment problem in HDCN, to handle the batched arrivals of communication flows. Their objective is to find, for each batch of flows, the corresponding hybrid routing paths.

In this chapter, we will review the different routing and wireless resource allocation strategies in HDCN. For the sake of completeness, we first give a brief description of the above problems and their challenges in HDCN. Then, in the second section, we will give an in-depth overview of the wireless channel allocation approaches dealing with one-hop inter-rack communications in HDCN. Next, we introduce the major joint online routing and channel allocation strategies for multi-hop communications in HDCN. Afterward, the main JBRC allocation algorithms dealing with the batched arrival of inter-rack flows are detailed. Then, we will present a qualitative comparison between the different related resource allocation and routing strategies in HDCN. Finally, we summarize this chapter.

3.1 Routing and Wireless Channel Allocation Problematic in HDCN

Intra-DCN communication flows can be either within the same rack (i.e. intra-rack) or between servers from different racks (i.e. inter-rack). In the context of hybrid (wired/wireless) data center networks (HDCN), augmented with inter-rack wireless links to alleviate over-subscription, researches mainly focus on inter-rack communications. An inter-rack communication request is characterized by: (i) a sending rack, (ii) a receiving rack, and (iii) a traffic flow to be transmitted between them. We recall that each top of rack ToR deploys: (i) a wireless transmission unit (WTU) which is equipped with four IEEE 802.11ad transceivers/antennas and (ii) a wired Ethernet switch. One of the key features of HDCN is its ability to efficiently: (i) allocate wireless/wired resources and (ii) route flows, for on-demand intra-DCN communications. In this respect, an efficient wireless channel allocation strategy is required so that both wireless and wired links in HDCN are judiciously allocated to ongoing communications while minimizing the end-to-end delay. The main objective of wireless channel allocation problem in HDCN is to maximize the proportion of intra-data center communication requests transiting over the wireless infrastructure. In doing so, the end-to-end delay of communications and the congestion of wired infrastructure are minimized, and hence, the total throughput in the HDCN is maximized. Formally, the main purpose is to satisfy each communication flow requirements, in terms of bandwidth, while minimizing congestion and alleviating interference between ongoing wireless links. It is worth noting that wireless channel allocation problem in HDCN has proven to be `NP-hard` (Cui et al., 2011c), due to interference constraint and the limited number of wireless channels (Halperin et al., 2011) (Zhang et al., 2011).

Furthermore, the hybrid DCN architecture is faced to the short range limitation of the 60GHz frequency band. Consequently, inter-rack communications cannot always be ensured in a single hop. To deal with this challenge, a few recent approaches have addressed the joint routing and channel allocation problem. The key insight of these methods is to jointly harness wireless and wired interfaces to enhance the data center network capabilities in term of bandwidth. In doing so, the end-to-end delay and the congestion of wired infrastructure are minimized. Formally, assuming an inter-rack communication flow from a source to a destination, the objective is to compute the best hybrid (i.e. formed by wireless and/or wired links) routing path while assigning wireless channels along links. The complexity of such a problem resides in the fact that channel allocation along the routing path should consider both: (i) the available bandwidth on each link and (ii) the level of wireless interference among intra-flow and inter-flow links, so that the end-to-end delay can be reduced.

3.1.1 Routing and Wireless Channel Assignment Challenges in HDCN

The routing and wireless channel allocation problem, for both one-hop and multi-hop communications, is extremely challenging for many reasons:

- *Arrival of inter-rack communication flows*: The inter-rack communication flows arrive to the HDCN is in dynamic way. Several research works Hamedazimi et al. (2014), Luo et al. (2016) model the arrival time of such requests as a Poisson process distribution with an inter-arrival λ_A. Each communication flow is characterized by its: (i) source rack, (ii) destination rack, (iii) arrival time, and (iv) volume of traffic. According to the distance between the sending and receiving racks, the flow can be transmitted either in one single hop or multiple hops. Communication flows are not predictable in advance, as they dynamically arrive to the data center and transmit a random traffic. Therefore, their processing is extremely hard since the traffic in real DCN environment is very unbalanced, while the response time should be minimized as long as possible.

- *Unbalanced traffic demands in HDCN*: One main specificity of traffic demands in data center applications is its unbalanced criteria. That makes, unfortunately, the resource management harder in HDCN. Indeed, traffic unbalancing entails traffic concentration problem. For instance, recent traffic statistics obtained from real DC applications such as map-reduce usually concentrate their traffic in only a few hot nodes (Cui et al., 2011c). The latter induces bottlenecks and further delay the completion time of ongoing communications. Moreover, the random distribution of hot nodes makes it challenging to properly add new wireless links and alleviate top of the racks (ToRs) congestion.
- *Wireless interference constraints*: Only four wireless channels are available for each deployed antenna operating with the IEEE 802.11ad standard. Although those channels are orthogonal and can be used simultaneously by the same rack, the traffic density in HDCN is likely to induce interference problem. In fact, wireless links that are in the same interference area cannot make use of the same wireless channel at the same time. Otherwise, collisions will occur in the medium and consequently the quality of solution (QoS) will be deteriorated. Therefore, wireless links should be appropriately established between ToRs in such a way that avoids interferences between wireless channels. It is worth noting that for the case of joint routing and channel assignment problem, two kinds of interference have to be considered. Actually, collisions may occur between links of the same routing path supporting the flow (intra-flow interference) as well as between links from different paths (i.e. flows) (inter-flow interference).
- *Limited resources*: Both wireless and wired resources in HDCN are limited. In fact, a single ToR switch is shared by all the servers of the same rack. Therefore, if a rack participates to many communications simultaneously, then the wired uplinks and downlinks of the ToR switch will be strongly congested. Moreover, only four wireless channels are available on each ToR. DC provider must optimize the allocation of wireless antennas and channels in aim to maximize the network performance.
- *it Congestion on ToR switches*: ToR switches suffer from high congestion level. Hotspot links are consequently emerging in the HDCN and over-subscription has to be alleviated by properly allocating non-interfering wireless links.
- *Decision-making*: The routing and wireless resource allocation in hybrid DCN can be performed either in a centralized or in a distributed way. In the centralized scheduling (Halperin et al., 2011, Cui et al., 2011c, 2013, Han et al., 2015), a single centralized controller in the DCN infrastructure is responsible for both the traffic collection and the decision processing. Specifically, having a global view on the available resources in the

HDCN, the centralized controller makes an optimal decision about the routing and resource allocation for the incoming traffic requests. Despite the advantages of such an approach, the centralized controller may be a bottleneck and a single point of failure. In the distributed decision (Cui et al., 2011b, Luo et al., 2016, Cui et al., 2011a), the routing and channel allocation decision is performed by different nodes in the DCN. Each entity has a local view of the DCN and is able to resolve a part of the decision problem. Then, all the decision makers coordinate together to find the global best solution. However, it is straightforward to see that there is no guarantee of the optimality.

3.1.2 Routing and Wireless Channel Assignment Criteria in HDCN

Both routing and channel allocation mechanisms should take into account several criteria related to the network performance and to the infrastructure provider revenue. Typically, the most relevant criteria considered in the context of HDCN consist in:

- *Network throughput*: The main objective of Cloud data center providers is to enhance network performance by maximizing the throughput of applications. Typically, the total network throughput corresponds to the cumulative transmission throughput of the traffic carried through the hybrid DCN.
- *Traffic volume*: Obviously, the total throughput is an important metric for wireless resource allocation problem. However, it is not sufficient in the context of HDCN. In fact, racks requesting a higher amount of traffic usually requires longer time to carry their transmission due to the bandwidth limitation. Thus, they are likely to further increase the global completion delay. Accordingly, traffic volume of communication flows strongly impact the HDCN performance and it is in general considered in related work such as Cui et al. (2011c).
- *Total network delay*: Estimating the network delay of each communication is mandatory to ensure a good DCN performance. In fact, a transmission with a high network delay that is caused by a congestion or a long communication path, may deteriorate DCN QoS. Thus, it is judicious to deploy wireless links in order to reduce the latency. The total delay of the network defines the cumulative transmission delay of all the finished communications in the network.
- *Spectrum spatial reuse* (SSR): Enhancing the spectrum reuse in very important to ensure an optimal use of the wireless infrastructure in

the HDCN. SSR of a channel corresponds to the number of wireless communications which are simultaneously using the same wireless channel. Note that four wireless channels are available for IEEE 802.11ad.

- *Link distance*: Corresponds to the distance between the two communicating servers or racks. Actually, each rack in the HDCN is defined with its geographical position, and accordingly the hop distance, between the source and the destination of each transmission, can be defined. The latter strongly impacts the network utility. In fact, flows with longer paths usually induce higher transmission latency and thus increase the load of switches. Therefore, it is usually recommended to assign such flows to wireless links so as to alleviate congestion. However, this solution may incur a higher potential interference on wireless links. Further, the distance between two communicating racks decides whether a single-hop or multi-hop communication has to be established. Authors of Cui et al. (2011b) consider this parameter to define their objective network function.

- *Interference rate*: The set of interfering links on an interface is a decisive parameter that impacts the quality of the link. In fact, the larger is the number of conflict edges, the higher the latency is, which may aggravate network performance.

- *Link cost*: The link cost is a crucial metric that deeply impacts the HDCN efficiency. In fact, it is an incarnation of the link congestion level, and the transmission delay. It is judicious to allocate wireless and/or wired links with lowest costs. It is worth pointing out that the cost of a link incarnates the transmission delay of its residual (wireless or wired) traffic and the resulting retransmission delays (wireless) caused by/on interfering links.

- *Wireless requests use*: To evaluate the ability of the wireless resource allocation and routing strategies to efficiently carry incoming communications while minimizing congestion, it is important to evaluate the rate of requests that are assigned to wireless channels. In doing so, the efficiency of decision algorithms in allocating resources is gauged.

3.2 Wireless Channel Allocation Strategies for One-Hop Communications in HDCN

We investigate, in this section, the existing wireless channel assignment approaches proposed for one-hop communications in hybrid DCNs. These strategies deal with wireless channel allocation problem for communications between two racks within the same transmission range. They can be classified into two main classes: (i) omni-directional antennas based strategies and (ii) beamforming-based strategies.

Hereafter, we will, first, discuss the main specificity distinguishing the wireless channel allocation problem in HDCN from that in classical wireless networks. Next, we will discuss in details the main proposals found in the literature.

3.2.1 Channel Allocation Problem in Wireless Networks

A rich panoply of researches have been studying the problem of channel allocation for wireless and cellular networks in the last decade. For instance, several approaches have been recently proposed to deal with this issue in the context of cellular mobile networks (Audhya et al., 2011, Mishra and Saxena, 2011). The main challenge in such a problem lies in ensuring an efficient utilization of channels while considering interference constraints. To do so, several heuristic techniques, such as genetic algorithm, tabu-search, and simulated annealing, have been used to tackle this NP-hard problem. In the other hand, wireless spectrum allocation has been addressed in the context of IEEE 802.11 wireless local area networks (WLANs) so as to judiciously assign channels among Access Points (Chakraborty et al., 2016). In addition, this issue was tackled for sensor networks as in Saifullah et al. (2014), Chowdhurya et al. (2009), by proposing efficient protocols for multi-channel communications for IEEE 802.15.4 wireless sensor network (WSN) while minimizing interference. It is worth pointing out that despite the performance of such proposed channel allocation solutions in the context of sensor or cellular networks, they cannot, unfortunately, be applied for HDCN. Actually, we harness in hybrid data centers both wireless and wired resources. In other words, not only interference constraints are taken into account but also the waiting delay on IP queues. Thus, both wired and wireless interfaces are jointly considered during the allocation process, in such a way that maximizes the amount of traffic transiting over wireless links, so that congestion on ToRs is alleviated and the throughput is enhanced.

3.2.2 Omni-Directional Antennas Based Strategies

- In Cui et al. (2011c), the authors propose a hybrid Ethernet/wireless DCN architecture to handle the limitations of Ethernet-based DCN architectures and boost network performance. The wireless channel allocation problem is formulated as an optimization problem where the objective is to maximize the total throughput while satisfying interference constraints. In this context, a Genetic heuristic-based approach, named Genetic-HDCN is put forward to solve the optimization problem while handling traffic demands. Formally, each individual is defined as the channel allocation scheme associating to each ongoing transmission link the proper channel. A feasible individual is a channel allocation scheme satisfying interference constraints. The individual candidates that have the highest total throughput are selected. Moreover, Genetic-HDCN makes use of improved crossover and mutation operators. However, the initial population of solutions is randomly generated by the Genetic algorithm which may notably affect the quality of the final solution. Furthermore, the proposed solution is heuristic based and hence does not guarantee an optimal or near-to-optimal solution. Moreover, it is well known that Genetic heuristic struggles to converge for some problem instances. According to the simulation results, Genetic-HDCN strategy improves the HDCN performance compared with the conventional Wired-DCN approach. Genetic-HDCN pseudo-code is summarized in Algorithm 3.1.

Algorithm 3.1: Genetic-HDCN pseudo-algorithm

---

1 Inputs: *m individuals* $X = \{X_1; X_2; \dots; X_m\}$

2 Output: *optimal solution* $Y = \{Y_1; Y_2; \dots; Y_m\}$

3 $Y \leftarrow \emptyset$

4 **while** *There is evolution for one generation* **do**

5     $X_1 \leftarrow$ Selection(X)

6   Divide the individuals in X_1 into pairs randomly; denote the set of pairs as X_p

7   Apply Crossover operator

8   Apply improved Mutation operator

9     $Y \leftarrow$ Individual with best fitness

- In Cui et al. (2013), the authors deal with the dynamic channel scheduling in wireless DCN. This approach assigns a weight to each edge. The latter corresponds to the transmission delay and reflects the level of the link contribution to the global DCN performance. The wireless transmission scheduling is formulated as an optimization problem. Then, a 0.5-approximation algorithm, Approximation-HDCN, is propounded to find an optimized channel allocation solution. This algorithm is based on a relaxation-rounding technique dealing with the relaxation of the original integer optimization problem. To prove its efficiency, the authors compare the performance of their approximation algorithm to their previous proposal Genetic-HDCN (Cui et al., 2011c). Simulation results show that this approach outperforms the heuristic-based solution, and both strategies improve the performance compared with Wired-DCN. Unfortunately, this paper assumes omni-directional antennas deployed in ToRs, which maximizes the interference effects in the HDCN.

- The authors of Cui et al. (2011b) consider each ToR as a WTU. They formulate, first, the one-hop channel allocation problem in HDCN as an optimization scheme while maximizing the utility of the network. Such a utility is defined as the product of the traffic amount transiting through the wireless infrastructure and the distance between the source and the destination. Then, they propose a heuristic approach based on Hungarian Algorithm, denoted by Hungarian-HDCN, to solve the problem. Typically, Hungarian-HDCN starts by defining a utility matrix U in which each entry corresponds to the utility of the link connecting two nodes in the network. Besides, Hungarian-HDCN repetitively performs Hungarian algorithm on U during each iteration, in order to compute the maximum weighted matching. The matching associates to each communication link the corresponding wireless or wired channel. At each iteration, the network utility is updated by subtracting the traffic from the new allocated links. The process is repeated until all the entries in the utility matrix become null, in which case all the wireless links are assigned to communications. Based on this approach, the best solution is greedily reached. Unlike the aforementioned work (Cui et al., 2011c, 2013), the authors assume that a wireless communication can be simultaneously transmitted through multiple links and adopt, hence, a dynamic programming approach to handle this distinction. It is worth noting that the channel allocation decision is made according to the already transmitted traffic which may affect the QoS in case of sporadic traffic demand. The pseudo-algorithm of Hungarian-HDCN is summarized through Algorithm 3.2.

Algorithm 3.2: Hungarian-HDCN pseudo-algorithm

---

1 Inputs: HDCN, m ongoing communications

2 Output: *optimal matching M*

3 $M \leftarrow \emptyset$

4 $U \leftarrow$ Compute Initial Utility Matrix

5 **while** $U \neq 0$ **do**

6     $M \leftarrow$ Compute-MaximumWeightedMatching-Hungarian

7   Set up links and allocate traffic

8     $U \leftarrow$ Update Utility Matrix

● In Cui et al. (2011a), a new wireless link scheduling in wireless DCN is propounded. It is worth noting that the scheduling corresponds to setting up wireless links so as to alleviate congestion on hot nodes, while properly allocate channels to avoid interference. Formally, the wireless scheduling problem is modeled using two optimization objectives. The first formulation is a Min–Max optimization problem that aims to minimize the maximum remaining utility (defined in Cui et al. (2011b)) after a transmission period while satisfying interference constraint. In doing so, the authors deal with the unbalanced traffic distribution. To solve the Min–Max problem, they propose a Greedy-based algorithm, named MM-Scheduling. Specifically, MM-Scheduling repetitively selects the hottest pending node v and seeks to allocate all the transmissions through v as long as a wireless link is available. The process is repeated until all pending nodes are allocated. MM-Scheduling is described in Algorithm 3.3.

Algorithm 3.3: MM-Scheduling pseudo-algorithm

---

1 Inputs: $\mathcal{G} = (\mathcal{V}, \mathcal{E})$, set of available channels C, set of traffic demand $T(\mathcal{E})$

2 Output: Channel allocation scheme S

3 $S \leftarrow 0$

4 $\mathcal{V}_p \leftarrow \mathcal{V}$

5 **while** $\mathcal{V}_p \neq \emptyset$ **do**

6     $v \leftarrow$ Select-Hotest-Pending-node

7   **if** *v has no available antenna **OR** has no remaining traffic* **then**

8        $\mathcal{V}_p \leftarrow \mathcal{V}_p - v$

9   **else**

10        $e \leftarrow$ Select a random transmission including v

11        $c \leftarrow$ Select a random available channel on e

12   Assign c for e

13        $S(e, c) \leftarrow 1$

14 return S

The second formulation aims to maximize the total network utility. The authors makes use of their previous heuristic-based approach Hungarian-HDCN (Cui et al., 2011b) to solve the best-effort optimization problem. Simulations results compare the effectiveness of the two approaches and show that MM-Scheduling outperforms Hungarian-HDCN in the case of uniform traffic distribution, while the two proposals reach similar results for hotspot traffic.

- In Shan et al. (2014), the authors conceive a hybrid DCN architecture based on Fat-Tree design. The main idea of their proposal is to combine wired and wireless links in the same communication path. Specifically, this approach aims at minimizing hotspots formation by proposing a new logical topology for the DCN that considers IP address assignment and traffic engineering scheme. However, we notice that, they only add the wireless links in the neighborhood of the source node, while wired links are used only to deliver traffic between relay ToRs leading to the final destination.

3.2.3 Beamforming-Based Strategies

Despite the undeniable success of 60GHz technique and its role in enhancing wireless DCN performance, it raises the challenge of the short transmission range. In this context, we noticed that a few recent approaches have explored the use of beamforming mechanism to carry direct inter-rack communication links in HDCN. Hereafter, we will discuss the main strategies deploying directional antennas for one-hop transmissions.

- In Zhang et al. (2011), the authors explore the feasibility of the 3D beamforming primitive in data centers. Based on experimental testbed design, they prove that this technique enhances wireless links capacity and further alleviates interference compared to 2D beamforming. Moreover, this approach augments the number of current wireless transmissions in the DCN. Specifically, they show that 3D beamforming technique eliminates link blockage thanks to the ceiling reflectors. Consequently, any two racks in the DCN can communicate directly with each others using only one hop link, without the need for routing. Nevertheless, this paper only focuses on studying the feasibility of 3D beamforming technique but does not address the wireless channel allocation issue in HDCN.
- In Zhou et al. (2012), the authors extend their prior work (Zhang et al., 2011) by further tackling the wireless channel assignment problem in HDCN. Basically, their purpose is to address the short range and link blockage limitations of the 60GHz technique by deploying 3D beamforming mechanism. The main contribution consists in building a small experimental testbed to prove the capacity of 3D beamforming to address the above challenges. Next, they propose a heuristic-based link scheduler algorithm, named `Greedy-HDCN` to allocate channels for ongoing communications. Typically, their proposal, makes use of a greedy heuristic so that the number of allocated concurrent links is maximized. To do so, the interference level of each link is estimated

by computing the predictable signal-to-noise-plus-interference ratio (*SINR*) (see Section 3.1) values on conflicting edges. Then, the graph coloring is performed on links in such a way that conflicting edges have to be colored with different colors (i.e. channels). The main idea of the Greedy-HDCN heuristic is to sort the edges according to their conflict degrees (i.e. number of non-scheduled interfering edges). Then, channels are allocated to links in a greedy fashion. This approach is processed in a centralized manner by the centralized controller of the HDCN. We summarize the proposal through the pseudo Algorithm 3.4.

Algorithm 3.4: Greedy-HDCN pseudo-algorithm

1   Inputs: $\mathcal{G} = (\mathcal{V}, \mathcal{E})$, set of available channels C
2   Output: Channel allocation scheme S
3   $S \leftarrow 0$
4   $L \leftarrow$ Set of non-scheduled ongoing communications
5   **while** $L \neq \emptyset$ **do**
6   Compute the conflict degree of each link in L
7   Sort the set of concurrent links according to the conflict degree
8      $e \leftarrow$ Link-With-Highest-Conflict-Degree
9      $c \leftarrow$ Allocate-Channel(e)
10     $S(e, c) \leftarrow 1$
11     $L \leftarrow L - e$
12   return S

However, we notice that Greedy-HDCN is non-preemptive, as it keeps unchanged the channels of ongoing communications. Moreover, it requires a mechanical rotation mechanism to frequently rotate antennas inducing, hence, an extra delay. Further, this approach is very specific to the 3D beamforming based HDCN, where each two racks can directly communicate in only one single link. In fact, using only small number of racks, mirrors are used to reflect signals between racks, so that to avoid multi-hop communications. However, this cannot be deployed for large scale DCNs due to the physical challenges and construction costs.

- In Vardhan et al. (2014), the authors propounded a new fully wireless DCN topology arranging all racks in a single hexagonal arrangement instead of the classical row one. They made use of IEEE 802.15.3.c standard (IEEE Std 802.15.3c-2009, 2009) to deploy the 60GHz wireless links. Not only this approach makes possible the communication between adjacent racks but also it enables communications between servers in the same rack, by adequately positioning the transceivers to form a polygon. Indeed, the authors enabled transceivers rotation (i.e. beam steering mechanism) in order to communicate with racks in different orientations via only point-to-point links. Note that this approach assumes that each rack has only two transceivers which limit the number of communications that can be simultaneously performed by a node. Moreover, since each node can communicate with only two neighbors simultaneously, multi-hop communications was not the prior focus of this paper. In fact, they only refer to a medium access control (MAC) layer mechanism (Singh et al., 2007) to deal with two-hop communications.

As a first contribution of this chapter, proposed a new wireless channel allocation mechanism for inter-rack communications in HDCN. Our approach, denoted by graph coloring-HDCN (GC-HDCN), leverages the wireless infrastructure in order to enhance network performance (Dab et al., 2015), (Dab et al., 2020). Unlike Cui et al. (2011b), we assume that the DCN traffic is unsplittable and hence carried through a single channel. Besides, while the channel scheduling mechanism in Cui et al. (2013) accords high priority to ongoing traffics, our proposal does not distinguish between incoming communications and aim to enhance the overall QoS required by applications. Moreover, contrarily to Zhang et al. (2011), Zhou et al. (2012), we assume both omni-directional and 2D directional antennas in order to avoid rotation delay induced by 3D beamforming transceivers. Moreover, we establish wireless links only between racks in the same transmission range. In doing so, we overcome the physical challenges of 3D beamforming technique that requires perfect ceiling positioning in DCNs.

3.3 Online Joint Routing and Wireless Channel Allocation Strategies in HDCN

Although 60GHz technique provides additional bandwidth to data center applications, prior proposals studied so far, restricted the wireless communications to the neighboring racks while carrying one-hop transmissions. This assumption dramatically limits the distance and the number of wireless links deployed in HDCN. Moreover, despite the ability of 3D beamforming to overcome short range limitation, it entails several physical challenges.

In this regard, a recent research approaches have dealt with multi-hop communications in HDCN. Although this issue has been heavily studied in the literature in the context of Mesh networks (Langar et al., 2010, Waharte et al., 2009, 2006, Bezahaf et al., 2012), the related approaches ensure only fully wireless paths which is unfortunately not applicable to HDCN.

In this section, we first summarize the most relevant related work in the context of Mesh network, that helped us to have an insight into joint routing and channel assignment in Hybrid DCN. Next, we review the main related strategies dealing with multi-hop communications in HDCN in online mode.

3.3.1 Joint Routing and Channel Assignment in Mesh Networks

- In Tang et al. (2005), the authors make use of multi-commodity flow model to deal with single joint routing and channel assignment in multi-channel wireless mesh networks. They aim to find the suitable routing path with the channel assignment for each communication while minimizing traffic effects. They propose a heuristic algorithm that succeed at solving the routing model in polynomial time. However, their proposal cannot be applied to a batched arrivals of requests since it accommodates only one single communication flow at once.

- In Kolar and Abu-Ghazaleh (2006), the authors address the same problem but for a batch of communication flows in multi-hop wireless networks. However, their approach does not ensure the channel assignment along the routing paths. Indeed, the authors seek to minimize the contention effects between the ongoing links, without prohibiting it.

- In Xiao et al. (2004), the authors tackle the problem of joint routing and resource allocation in wireless data networks. They formulate the problem based on an integer linear programming (ILP) statement, and make use of a dual decomposition method to solve it. This approach does not take into consideration interference constraint in the routing path. Moreover, it enables completely wireless communication routes, which is not always the case for HDCN architecture.

- A rich research work as in Islam et al. (2015), Mohsenian-Rad and Wong (2007), have reviewed the joint routing and channel assignment in multi-channel wireless mesh networks.

Unfortunately, these mechanisms cannot be applied in the context of HDCN, where both wireless and wired interfaces have to be considered. Moreover, in HDCN, additional constraints have to be considered during the decision process. Namely, wireless interferences and the length of IP queues (waiting delay) should be jointly optimized to enhance the routing of communication flows.

3.3.2 Online Joint Routing and Channel Assignment Strategies in HDCN

The joint routing and channel assignment strategies in HDCN provide the hybrid (wireless/wired) routing path for each single incoming communication request, in an online way. Hereafter, we will review the main relevant strategies found in the literature.

- In Halperin et al. (2011), the authors propound a new augmented data center architecture by deploying the 60GHz wireless technology in their proposed VL2 architecture. A Greedy-Flyway-HDCN strategy is proposed and greedily augments the wired DCN with extra flyways. The latter are 60GHz wireless links which are set up between top-of-rack switches as long as there is network congestion. In doing so, bandwidth capacity is increased. Note that each flyway is considered as (i) one-hop wireless communication and (ii) not involved in the routing process. If the state of wired network is not loaded, wired infrastructure VL2 routes the traffic using wired link-state IP routing, open shortest path first (OSPF), and equal-cost multi-path (ECMP) protocols. In the case of congestion, a flyway is setup and the appropriate route is statically updated at the ToR so that the traffic passes through the wireless links. Note that each flow must transit through exactly one flyway. Greedy-Flyway-HDCN focuses on alleviating congestion effects by statically including flyways in wired routing paths. In other words, the proposal deals only with hotspots links. Unfortunately, the wireless channel allocation and wireless multi-hop are not considered since only non-interfering flyways are greedily added. The pseudo-algorithm of Greedy-Flyway-HDCN is summarized in Algorithm 3.5.
- In Shin et al. (2013), the authors propose a fully wireless data center architecture named Cayley data center topology. Racks are attached to densely wireless connected mesh topology in aim to maximize the number of active wireless links. In order to alleviate interference effects, this strategy makes use of beamforming technique with fixed-direction antennas. The routing is based on a geographic approach, denoted XYZ-Routing, which finds the intra and/or inter rack path. In fact, the next hop server is the closest one to the final destination. We notice that the authors focus only on minimizing the routing path length. Indeed, the routing decision only depends on the geographic position of the destination. In doing so, some wireless links may be excessively used and induces high probability of collisions which mitigates network performance. Moreover, this strategy does not consider wireless channel

---

Algorithm 3.5: `Greedy-Flyway-HDCN` pseudo-algorithm

---

1 Inputs: HDCN, set of available channels C, set of communications C

2 Output: $\mathcal{F}$ Flyway links

3 $\mathcal{F} \leftarrow \emptyset$

4 $H \leftarrow$ Set of Hotspot links

5 **while** $H \neq \emptyset$ **do**

6     $h \leftarrow$ Select-Hotspot

7   **if** *HotSpot-On-Source* **then**

8         $f \leftarrow$ Choose-Flyway-From-Source

9   Allocate-Channel-ToFlyway(f)

10   **else**

11   /*Flyway in Destination*/

12         $f \leftarrow$ Choose-Flyway-To-Destination

13   Allocate-Channel-To-Flyway(f)

14     $\mathcal{F} \leftarrow \mathcal{F} \cup f$

15   Construct-Routing-Path(f)

16     $H \leftarrow H \setminus h$

17 return $\mathcal{F}$

---

allocating jointly to the routing process. Instead, wireless channels are arbitrated based on a MAC layer arbitration protocol along the path.

The geographical routing protocol `XYZ-Routing` is summarized in Algorithm 3.6.

- In Li et al. (2014), the authors propose spherical mesh topology for wireless DCN. The racks within the same wireless transmission range are regrouped into a spherical unit. The main idea is to take profit of the geometric characteristics of the spheres to eliminate link congestion by placing antennas over them. The routing algorithm, named `Spherical-HDCN`, is based on geographical approach that gets the route depending on the position of the spheres containing the two communicating servers. Unfortunately, we notice that this strategy is very specific for the above particular spherical topology and cannot be applied to the common DCN architectures. Moreover, the proposal does not take into consideration channel assignment along the routing path.

- In Zhu et al. (2014), the authors explore the wireless infrastructure only for the control plane while data is completely transiting over wired infrastructure. The objective is to ensure a highly available control functions

---

Algorithm 3.6: XYZ-Routing pseudo-algorithm

1 Inputs: Cayley HDCN, communication C, g_{src}

2 Output: routing path $\mathcal{P}$

3 g_{curr} ←geographical position of the server containing current pacet

4 r_{curr} ←rack of the current server

5 g_{dst} ←geographical position of the final destination

6 r_{dst} ←rack of the final destination

7 $\mathcal{R}_{adj}$ ← Set of racks adjacent to r_{curr}

8 $g_{curr} \leftarrow g_{src}, \mathcal{P} \leftarrow g_{curr}$

9 **while** $g_{curr} \neq g_{dst}$ **do**

10   **if** *IsInDifferentRack(g_{curr}, g_{dst})* **then**

11   r_{next} ← Get-Min-Distance-Rack(r_{dst}, $\mathcal{R}_{adj}$)

12   **else**

13   /*same rack but different servers*/

14   g_{next} ← Get-Min-Distance-Rack(g_{curr}, g_{dst})

15      $\mathcal{P} \leftarrow \mathcal{P} \cup g_{curr}$

16 return $\mathcal{P}$

---

by alleviating interference effects and enhancing the throughput. To do so, 3D beamforming using horn/array antennas with static directions are deployed. Note that the calibration of directions aims to minimize the inter-flow interferences. In addition, new routing algorithm based on Kautz graph is proposed for signalization traffic. The key idea of this algorithm is to seek for the shortest path. Unfortunately, wireless channels over the routing path are assigned based on a simple greedy heuristic that minimizes intra-path interference but does not nullify it. Besides, the use of static 3D antennas direction strongly limits the usage of spectrum. Finally, this strategy only investigates the wireless links in the control plane, and does profit from this promising technology to alleviate massive traffic explosion in the data plane. Therefore, the proposed routing approach cannot be applied to deal with inter-rack communication in modern HDCNs.

- In Hamedazimi et al. (2014), the authors make use of free-space optical technique to augment data center network with wireless links. The wireless links are established by deploying mirrors and lens on ToRs. Note that their optical architecture ensures free-interference wireless communication links. They formulate the routing problem using the maximum weighted matching and solve it based on a heuristic selecting minimum hop-count alternating paths. Nevertheless, this approach only considers the hop count during the routing process, since the optical technique does not require the wireless channel assignment along the path. In doing so, several important network metrics are neglected, such as the waiting delay in IP queues, link congestion, etc.

- In Han et al. (2016), the authors investigate, from a cross-layer view, the use of wireless infrastructure to augment the wired DCN so that to alleviate link over-subscription. This strategy separately tackles the routing and wireless channel allocation problem. In fact, first, a routing protocol is proposed to minimize the hop counts of the routing flow path. The main idea is to establish wireless links only if they reduce the total number of hops. Besides, the authors deal with congestion problem by proposing an online wireless channel and power allocation algorithm. Indeed, contrarily to most of research works dealing with HDCN, they assume that the transmission power of wireless antennas is not fixed, and propose, hence, a Greedy-based heuristic to repetitively allocate the channel ensuring the maximum capacity gain. It is worth noting that this approach may not be efficient as it computes first the shortest routing paths without considering potential channel allocation. In fact, addressing jointly the two problems is more likely to optimize the wireless resource usage. Moreover, the proposals are validated for a small instance of DCN, composed by only 20 racks, and their efficiency for large-scale DCN is not guaranteed.

As a second contribution of this chapter, we propose a new online joint routing and channel assignment approach in HDCN, for inter-rack communications, while making use of 2D beamforming technique. Unlike Li et al. (2014), we assume common hybrid data center network architecture based on the well-known CLOS design, and our approach is not specific to a particular topology. Moreover, we do not assume static antennas' directions as in Shin et al. (2013), so that we maximize the usage of wireless interfaces. To overcome the rotation delay induced by horn antennas in Zhu et al. (2014), we make use of 2D switched beam antennas. Unlike Hamedazimi et al. (2014), Han et al. (2016), we take into account interference constraints during the routing decision. Indeed, it is not only the hop count that is considered during the path computation, but also other cost metrics. Our approach favors the paths that ensure the higher throughput by reducing interference effects. Unlike Shin et al. (2013), we pay attention to the link state during routing decision by favoring both wireless and wired interfaces with higher residual bandwidth in aim to enhance network performance. Hence, each routing communication path may be composed of wireless and/or wired links. Further, we deploy IEEE 802.11ad (IEEE

Std 802.11ad 2012, 2012) to build 60GHz wireless infrastructure instead of IEEE 802.15.3.c. standard, deployed in Shin et al. (2013). In fact, IEEE 802.11ad is better in terms of bandwidth and number of available channels. Finally, unlike Halperin et al. (2011), each routing communication path may be composed of wireless and/or wired links.

3.4 Joint Batch Routing and Channel Allocation Strategies in HDCN

While the above related strategies process each single communication flow in an online way, few recent research approaches have dealt with the problem in a batch mode. The main objective of such a mode is to handle the unbalanced and heavy traffic, by carrying the batched arrivals of communication flows, and hence to ensure a better use of HDCN resources. In doing so, the communications, arriving during a specific time window, are queued together and their processing is delayed to the following time window.

3.5 Joint Batch Routing and Channel Allocation Strategies in HDCN

While the above related strategies process each single communication flow in an online way, few recent research approaches have dealt with the problem in a batch mode. The main objective of such a mode is to handle the unbalanced and heavy traffic, by carrying the batched arrivals of communication flows, and hence to ensure a better use of HDCN resources. In doing so, the communications, arriving during a specific time window, are queued together and their processing is delayed to the following time window.

Note that, there is a variety of research work addressing the JBRC allocation in wireless mesh networks, as in Islam et al. (2015), Mohsenian-Rad and Wong (2007), and Kolar and Abu-Ghazaleh (2006). Unfortunately, the latter mechanisms are different from our problem (HDCN), where both wireless and wired interfaces must be considered.

Hereafter, we will discuss the main few research strategies dealing with the JBRC assignment in HDCN.

- In Han et al. (2015), the authors propose a RUSH framework for joint: (i) routing and (ii) scheduling wireless antennas in HDCN in both online and batch modes. They design a three-layer multi-rooted DCN topology where each rack is equipped with only one 60GHz steerable directional antenna. Specifically, one antenna may be involved in many routing paths simultaneously. To do so, RUSH allocates non-overlapping time slots for different links, while minimizing the congestion load in the HDCN. The joint routing and scheduling problem in HDCN (JRSH) is formulated as an ILP model, and has as objective to minimize the maximum link congestion. In batch mode, RUSH framework makes use of RUSH-batch algorithm. The main idea of the latter is to relax JRSH problem and then solve it using an linear programming (LP) solver. RUSH-batch makes use of the LP fractional solution to randomly choose routing paths for each request. Besides, based on the congestion level on each path, a feasible antenna scheduling along the path is determined. In the online mode, the authors put forward a RUSH-online algorithm that sequentially computes the single shortest routing path while scheduling time slots. Note that RUSH strategy deploys beam steering to change the antenna direction during each time fraction, which may induce extra delays. The pseudo-code of the batch algorithm of RUSH framework is summarized in Algorithm 3.7.

Algorithm 3.7: RUSH pseudo-algorithm

Inputs: Request set $\mathcal{R}$, the solution to the LP-relaxation of JRSH
Output: Routing scheduled paths $\mathcal{P}$
$i \leftarrow 0$
for all request r_i in $\mathcal{R}$ **do**
  for all link e transmitting flow **do**
    Find the single path from s_i to d_i through e with minimum congestion load
  end for
  $p_i \leftarrow$ Pick a path
  Find a feasible scheduling on $\mathcal{P}$
  $\mathcal{P} \leftarrow \mathcal{P} \cup p_i$
end for

The same RUSH mechanism was used by the authors of Zhang et al. (2016), to find the hybrid routing path in the HDCN after a virtual machine deployment in the racks.

- In Luo et al. (2016), the authors propound a new DCN architecture, VLC-cube, by augmenting the Fat-Tree topology with optical wireless infrastructure. Specifically, all inter-rack communications are carried on only wireless links, using the visible light communication (VLC) techniques. The authors propose a new routing scheme that greedily seeks for the least congested hybrid path for each flow in both online and batch mode. Note that the proposed approach is very specific to VLCcube topology, since path computation depends on both the rack and pod placement. Moreover, the strategy only deals with routing problem regardless interference constraints and channel allocation problem since optical wireless communications are deployed.

- In He and Mao (2016), the authors deal with dual-hop routing for a set of communications requests (i.e. batch mode), in wireless dual-hop networks based on 60GHz. Typically, they always assume a two-hop networks where the hop count in the network can at most be equal to 2. The authors propound a decomposition heuristic method, Dual-Heuristic, to jointly optimize relay and link selection. The main objective of this strategy is to minimize the maximum expected delivery time. To do so, Dual-Heuristic decomposes, first, the original problem into a: (i) relay selection and (ii) link selection sub-problems, then, it develops a Greedy heuristic to alleviate time complexity. Note, however, that is approach is very restricted to a specific configuration where 60GHz wireless technique is used only for two hops, and cannot be applied in the context of HDCN. Moreover, it does not deal with channel assigning alongside the routing process.

In Dab et al. (2017b), Dab et al. (2017a), we proposed new JBRC assignment approach in HDCN, to deal with the batched arrivals of communication flows. It is worth pointing out that none of the previous strategies address the channel allocation jointly to the routing process in batch mode. Contrarily to Han et al. (2015), our approach deals with a batch of flows while allocating wireless channels along the paths. Moreover, unlike Luo et al. (2016), we design a hybrid DCN by augmenting the wired network with wireless communication links, and our proposal is generic and is not specific to a particular HDCN topology. Contrary to He and Mao (2016), our proposed algorithm does not limit the number of wireless links in the hybrid routing path. Finally, in Dab et al. (2017b), we validated the proposal based on realistic Facebook's traces. The results obtained show clearly the improvement in terms of QoS performance compared with related work.

3.6 Summary

Table 3.1 summarize a comparison between the aforementioned strategies for: (i) wireless channel allocation and (ii) online and batch joint routing and channel assignment, in HDCN. Specifically, we classify the related

Table 3.1 Summary of routing and channel allocation strategies in HDCN.

Strategy	Architecture	Solution	One/ multi hop	Mode	Constraints	HDCN technique	Wireless technique
Genetic-HDCN Cui et al. (2011c)	Tree-layered	Heuristic	One-hop	Online	Interference	Wireless/wired	Omni-directional 60GHz
Approximation-HDCN Cui et al. (2013)	Tree-layered	Approx	One-hop	Online	Interference	Wireless/wired	Omni-directional 60GHz
Hungarian-HDCN Cui et al. (2011b)	Tree-layered	Heuristic	One-hop	Online	Interference	Wireless/wired	Omni-directional 60GHz
MM-scheduling Cui et al. (2011a)	Tree-layered	Heuristic	One-hop	Online	Interference	Wireless/wired	Omni-directional 60GHz
Shan et al. (2014)	Fat-Tree	Heuristic	One-hop	Online	Interference	Wireless/wired	Omni-directional
Zhang et al. (2011)	Tree-based	—	One-hop	Online	Interference	Wireless/wired	3D beamforming
Greedy-HDCN Zhou et al. (2012)	Tree-based	Heuristic	One-hop	Online	Interference	Wireless/wired	3D beamforming
Vardhan et al. (2014)	Hexagonal Fat-Tree	—	One-hop	Online	Interference	Fully wireless	Beamforming 60GHz

Greedy-Flyway-VL2 HDCN Halperin et al. (2011)	Greedy	Routing	Online	Interference	Wireless/wired	Beamforming 60GHz
XYZ-Routing Cayley DCN Shin et al. (2013)	Geographic	Routing	Online	Path length	Wireless	Beamforming 60GHz
Spherical-HDCN spherical DCN Li et al. (2014)	Geographic	Routing	Online	Distance	Wireless	Beamforming 60GHz
Angora Zhu et al. (2014)	Kautz-graph	Routing	Online	Hop number	Wireless/wired	3D beamforming
FireFly Hamedazimi et al. (2014)	Heuristic	Routing	Online	Hop-count	Wireless optics	Free-space optics
Grid Han et al. (2016)	Decomposition method	Routing	Online	Power	Wireless	60GHz
Three-layered RUSH Han et al. (2015)	ILP relaxation	Routing	Online/batch	Time scheduling	Wireless/wired	Beamforming 60GHz
VLCcube Luo et al. (2016) Fat-Tree	Heuristic	Routing	Batch	Rack placement	Wireless optics	VLC
Dual-Heuristic He and Mao (2016) Dual-hop network	Heuristic	Two-hop routing	Batch	Delivery time	Wireless/wired	Beamforming 60GHz

method according to the: (i) deployed architecture, (ii) addressed problem (i.e. one-hop or multi-hop communications), (iii) processing mode (i.e. online or batch), (iv) constraints considering during the decision, and (v) deployed technique.

3.7 Conclusion

In this chapter, we provided a detailed overview of routing and channel allocation strategies in HDCN, for both one-hop and multi-hop inter-rack communications. First, we briefly described the wireless channel allocation problem for intra-DCN flows in single hop, and the joint routing and wireless channel assignment problem for multi-hop communications. Then, we addressed the main challenges encountered by this issue in HDCN. Afterward, we highlighted the most important criteria that have been considered when dealing with the routing and wireless channel allocation problems in HDCN. Next, we detailed the main related strategies that we classify into three main groups: (i) wireless channel allocation approaches dealing with one-hop communications in HDCN, (ii) online joint routing and wireless channel allocation approaches addressing multi-hop communications in HDCN in a sequential way, and (iii) batch joint routing and wireless channel assignment approaches handling the batched arrivals of communication flows to HDCN. Finally, we summarized the review with a qualitative comparison of the different proposed strategies.

References

G.K. Audhya, K. Sinha, S.C. Ghosh, and B.P. Sinha. A survey on the channel assignment problem in wireless networks. *Wireless Communications and Mobile Computing.* 11: 583–609, May 2011.

M. Bezahaf, L. Iannone, M.D. Amorim, and S. Fdida. An experimental evaluation of cross-layer routing in a wireless mesh backbone. *Elsevier Computer Networks Journal.* 56: 1568–1583, January 2012.

S. Chakraborty, B. Saha, and S.K. Bandyopadhyay. Dynamic channel allocation in IEEE 802.11 networks. *International Journal of Computer Applications.* 149: 36–38, September 2016.

K.R. Chowdhurya, N. Nandirajub, P. Chandac, D.P. Agrawald, and Q. Zenge. Channel allocation and medium access control for wireless sensor networks. *Elsevier Ad Hoc Networks.* 7: 307–321, March 2009.

Y. Cui, H. Wang, and X. Cheng. Wireless link scheduling for data center networks. *5th International Conference on Ubiquitous Information Management and Communication (ICUIMC'11)*, February 2011a.

Y. Cui, H. Wang, X. Cheng, and B. Chen. Wireless data center networking. *IEEE Wireless Communications*. 18: 46–53, December 2011b.

Y. Cui, H. Wang, and X. Cheng. Channel allocation in wireless data center networks. *IEEE International Conference on Computer Communications (INFOCOM)*, April 2011c.

Y. Cui, H. Wang, X. Cheng, D. Li, and A. Yla-Jaaski. Dynamic scheduling for wireless data center networks. *IEEE Transactions on Parallel and Distributed Systems*. 24: 2365–2374, December 2013.

B. Dab, I. Fajjari, and N. Aitsaadi. A novel joint routing and channel allocation approach in hybrid data center network. *IEEE International Conference on Sensing, Communication, and Networking (SECON)*, August 2017a.

B. Dab, I. Fajjari, and N. Aitsaadi. Online-batch joint routing and channel allocation for hybrid data center networks. *IEEE Transactions on Network and Service Management*. 14: 831–846, August 2017b.

B. Dab, I. Fajjari, N. Aitsaadi, and A. Mellouk. A novel wireless resource allocation algorithm in hybrid data center networks. *2015 IEEE 12th International Conference on Mobile Ad Hoc and Sensor Systems*, Dallas, TX, 2015, pp. 109–117. doi: https://doi.org/10.1109/MASS.2015.19.

B. Dab, I. Fajjari, and N. Aitsaadi. Optimized wireless channel allocation in hybrid data center network based on IEEE 802.11ad. *Computer Communications*. 160: 534–546, July 2020.

D. Halperin, S. Kandula, J. Padhye, P. Bahl, and D. Wetherall. Augmenting data center networks with multi-gigabit wireless links. *ACM Special Interest Group on Data Communication (SIGCOMM)*, August 2011.

N. Hamedazimi, Z. Qazi, H. Gupta, V. Sekar, S.R. Das, J.P. Longtin, H. Shah, and A. Tanwer. FireFly: a reconfigurable wireless data center fabric using free-space optics. *ACM Special Interest Group on Data Communication (SIGCOMM)*, August 2014.

K. Han, Z. Hu, J. Luo, and L. Xiang. RUSH: routing and scheduling for hybrid data center networks. *IEEE Conference on Computer Communications (INFOCOM)*, August 2015.

Z. Han, Y. Li, H. Tany, R. Wangz, and Y. Zhang. Cross-layer protocol design for wireless communication in hybrid data center networks. *The 12th International Conference on Mobile Ad-Hoc and Sensor Networks (MSN)*, December 2016.

Z. He and S. Mao. A decomposition principle for link and relay selection in dual-hop 60 GHz networks. *IEEE INFOCOM*, October 2016.

IEEE Std 802.15.3c-2009. IEEE Standard for information technology - telecommunications and information exchange between systems - local and metropolitan area networks - specific requirements. Part 15.3: Wireless Medium Access Control (MAC) and Physical Layer (PHY) Specifications for high rate wireless personal area networks (WPANs) amendment 2: Millimeter-wave-based alternative physical layer extension. *(Amendment to IEEE Std 802.15.3-2003)*, December 2009.

IEEE Std 802.11ad 2012. IEEE Standard for Information technology-Telecommunications and information exchange between systems-Local and metropolitan area networks-Specific requirements-Part 11: Wireless LAN Medium Access Control (MAC) and physical layer (PHY) Specifications Amendment 3: Enhancements for Very High Throughput in the 60GHz Band. *IEEE Std 802.11ad 2012*, December 2012.

A. Islam, Md.J. Islam, N. Nurain, and V. Raghunathan. Channel assignment techniques for multi-radio wireless mesh networks: a survey. *IEEE Communications Surveys and Tutorials*. 18: 988–1017, December 2015.

V. Kolar and N.B. Abu-Ghazaleh. A multi-commodity flow approach for globally aware routing in multi-hop wireless networks. *IEEE International Conference on Pervasive Computing and Communications (PERCOM)*, March 2006.

R. Langar, N. Bouabdallah, R. Boutaba, and G. Pujolle. Interferer link-aware routing in wireless mesh networks. *IEEE ICC*, May 2010.

Y. Li, F. Wu, X. Gao, and G. Chen. SphericalMesh: a novel and flexible network topology for 60GHz-based wireless data centers. *IEEE/CIC International Conference on Communications (ICCC)*, October 2014.

L. Luo, D. Guo, J. Wu, S. Rajbhandari, T. Chen, and X. Luo. VLCcube: a VLC enabled hybrid network structure for data centers. *IEEE Transactions on Parallel and Distributed Systems*. 28: 2088–2102, December 2016.

M.P. Mishra and P.C. Saxena. Issues, challenges and problems in channel allocation in cellular system. *Computer and Communication Technology (ICCCT)*, November 2011.

A.H. Mohsenian-Rad and V.W.S. Wong. Joint logical topology design, interface assignment, channel allocation, and routing for multi-channel wireless mesh networks. *IEEE Transactions on Wireless Communications*. 6: 4432–4440, December 2007.

A. Saifullah, Y. Xu, C. Lu, and Y. Chen. Distributed channel allocation protocols for wireless sensor networks. *IEEE Transactions On Parallel and Distributed Systems*. 25: 2264–2274, September 2014.

L. Shan, C. Zhao, X. Tian, Y. Chengy, F. Yang, and X. Gan. Relieving hotspots in data center networks with wireless neighborways. *IEEE Global Telecommunications Conference (GLOBECOM)*, December 2014.

J.Y. Shin, E.G. Sirer, H. Weatherspoon, and D. Kirovski. On the feasibility of completely wireless datacenters. *IEEE/ACM Transactions on Networking (TON)*. 21: 1666–1679, October 2013.

S. Singh, F. Ziliottoy, U. Madhow, E.M. Belding-Royerz, and M.J.W. Rodwell. Millimeter wave WPAN: cross-layer modeling and multihop architecture. *IEEE International Conference on Computer Communications, InfoCom*, May 2007.

J. Tang, G. Xue, and W. Zhang. Interference-aware topology control and QoS routing in multi-channel wireless mesh networks. *The ACM International Symposium on Mobile Ad Hoc Networking and Computing (MobiHoc)*, July 2005.

H. Vardhan, S.R. Ryu, B. Banerjee, and R. Prakash. 60 GHz wireless links in data center networks. *Computer Networks: The International Journal of Computer and Telecommunications Networking*. 58: 192–205, January 2014.

S. Waharte, R. Boutaba, Y. Iraqi, and B. Ishibashi. Routing protocols in wireless mesh networks: challenges and design considerations. *Multimedia Tools and Applications*. 29: 285–303, June 2006.

S. Waharte, B. Ishibashi, R. Boutaba, and D.E. Meddour. Design and performance evaluation of IAR: interference-aware routing metric for wireless mesh networks. *Mobile Networks and Applications (MONET)*. 14: 649–660, December 2009.

L. Xiao, M. Johansson, and S.P. Boyd. Simultaneous routing and resource allocation via dual decomposition. *IEEE Transactions on Communications*. 52: 1136–1144, July 2004.

W. Zhang, X. Zhou, L. Yang, Z. Zhang, B.Y. Zhao, and H. Zheng. 3D beamforming for wireless data centers. *HotNets*, November 2011.

W. Zhang, S. Zhang, Z. Qian, K. Wen, and S. Lu. Virtual network deployment in hybrid data center networks. *IEEE Trustcom/BigDataSE/ISPA*, August 2016.

X. Zhou, Z. Zhang, Y. Zhu, Y. Li, S. Kumar, A. Vahdat, B.Y. Zhao, and H. Zheng. Mirror mirror on the ceiling: flexible wireless links for data centers. *ACM SIGCOMM*, August 2012.

Y. Zhu, X. Zhou, Z. Zhang, L. Zhou, A. Vahdat, B.Y. Zhao, and H. Zheng. Cutting the cord: a robust wireless facilities network for data centers. *ACM Conference on Mobile Computing and Networking (MobiCom)*, September 2014.

4

Inter-Data Center Networks: Routing and Reliability in Virtual Network Backbone

Oussama Soualah[1,2], Ilhem Fajjari[3], and Nadjib Aitsaadi[4]

[1] OS-Consulting, Athis-Mons, Essonne, France
[2] LiSSi Lab, UPEC, Vitry sur Seine, Val de Marne, France
[3] Orange Labs, Orange, Chatillon, Hauts-de-Seine, France
[4] Universités Paris-Saclay, UVSQ, DAVID, F-78035, Versailles, France

Abstract

Few survivable virtual network (VN) embedding algorithms have been proposed in the literature. Indeed, we can classify the related methods into two main groups: (i) *centralized* and (ii) *distributed* approaches. Besides, in each group, the reliable VN embedding strategies assume that failures can occur in (i) routers and/or links failures in the whole $\mathcal{SN}$ or (ii) in a defined geographic region. Hereafter, we will present a summary of the prominent survivable strategies with respect to more criteria. To do so, first of all, we start by presenting a short overview of the baseline strategies which do not consider the reliability of physical resources. Afterward, we will provide a deep overview of the survivable VN mapping strategies.

4.1 Overview of Basic Virtual Network Embedding Without Reliability Constraint

The virtual network (VN) embedding problem is NP-hard (Kleinberg, 1996, Kolliopoulos and Stein, 1997). In order to skirt its complexity, IT researchers consider sometimes an infinite substrate resources or relax some other conditions. Besides, they opt to heuristic approaches to solve the VN mapping problem since the optimal solution is computationally intractable in large-scale networks. Hereinafter, we will summarize the most prominent online and batch algorithms.

4.1.1 Online Approaches

In Chowdhury et al. (2009), the authors propounded an original strategies that coordinate the virtual router and link embedding stages. The simultaneous combination of router and link mapping is a major pros of this work. The authors enhance the substrate network by fictitious resources called meta-nodes and meta-links. A meta node is created to match with the potential substrate routers that can host the current virtual router. Then, the connection is ensured by establishing meta-links between the meta-node and its potential routers. It is noteworthy that the meta-link capability is infinite. Later on, the authors formulate the embedding problem as mixed-integer linear programming (MILP). Basing on the above formalization, an optimized solution is reached after a constraint relaxation and two variants are proposed: (i) deterministic-ViNE (D-VINE) and (ii) randomized-ViNE (R-VINE). We notice that the choice of the potential substrate routers is based on geographic criterium; however, this can be a limitation to reach a pertinent solution. Besides, proposing a splittable approaches can be supported by the network provider only in the future because the current routers do not implement such kind of splittable behavior.

In Zhou et al. (2010), Wang et al. (2012), the authors formulated the VN embedding problem as a non-cooperative game. In the proposed mapping game, each one of the $\mathcal{VN}$ is represented by a decision maker (i.e. player). These players behave selfishly as competitors to satisfy their required bandwidth from the $\mathcal{SN}$ resources. In these papers, the authors did not justify why they devise a non-cooperate game. Furthermore, since the mapping algorithm will be executed by the $\mathcal{CP}$, the proposed solutions should enhance the revenue. In the proposed algorithms, the authors did not explain how may this non-cooperative game improve $\mathcal{CP}$ turnover. Besides, we notice that the authors did not deal with (i) possible failures within the $\mathcal{SN}$ and (ii) the substrate and virtual router capacities (i.e. memory and processing power). In fact, they only focused on the link resource (i.e. bandwidth) allocation. Moreover, the performance evaluation of the new proposal has been performed in a small-size $\mathcal{SN}$ which is not realistic. Accordingly, the authors did not study the proposal scalability. In fact, in this kind of small-size networks, the optimal solution can be computed in polynomial time without any optimized solution. Thus, the added value of the paper is not well defended. We notice, also, that the rejection rate of VN requests was not estimated in simulation scenario. Indeed, this latter metric is very important to describe the provider's turnover. Finally, the authors did not compare the propounded strategy with regard to the prominent related algorithms.

In Fajjari et al. (2011a), Fajjari et al. (2014), the authors propounded a new strategy, basing on the Ant Colony metaheuristic (Stützle and Hoos, 2000, Dorigo and Blum, 2005), named VNE-AC to tackle the high complexity of the VN embedding problem by considering the mapping of virtual links in unsplittable manner. The authors decomposed the VN in elementary topologies in order to apply the divide and conquer principle. The main objective of the authors is to maximize the Cloud provider's revenue by accepting the maximum of clients. However, this proposal suffers from the lack of an optimal network load balancing. In order to overcome the problem of the overloaded substrate links, the authors proposed a new strategy named virtual network reconfiguration (VNR) (Fajjari et al., 2011b) that reconfigures the network. It is noteworthy that this approach is a hybrid (i.e. reactive and proactive) strategy. VNR relies on the migration of some mapped virtual routers and their attached virtual links to other physical equipment in order to load balance the substrate network. By doing so, the Cloud's backbone would be able to host more clients. Accordingly, the provider turnover will be improved. Later on, in order to avoid the drawbacks of a static approach, the authors propounded an adaptive resource allocation scheme denoted Adaptive-VNE (Fajjari et al., 2012). This new strategy is based on the K-supplier method to embed the elementary topologies of a VN request. It is worth pointing out that the authors did not consider the mapping of virtual links in splittable manner in order to gauge their strategies with this class of algorithms.

4.1.2 Batch Approaches

Few methods tackled the problem of $\mathcal{VN}$ mapping in batch mode. Hereafter, we will describe some of the most prominent ones.

In Yu et al. (2008), the authors propounded a two-stages algorithm denoted by Batch-Baseline (B-Base) dealing with $\mathcal{VN}$s embedding in batch mode. B-Base sorts all queued $\mathcal{VN}$ requests with respect to their revenues and handles their mapping on this order. In the first stage, B-Base maps only virtual routers for the incoming $\mathcal{VN}$ requests. Then, the virtual links are embedded based on an unsplittable approach. We notice that B-Base adopts a greedy strategy for virtual routers and link mapping. In fact, this kind of approach may converge to a local optimum which will impact the CP's revenue. On the contrary, CP can earn more by accepting other VN requests with less revenue but the cumulative revenue would be higher than the one generated by accepting the requests having the maximum revenue. Moreover, the survivability of VN embedding is not considered.

In Jarray and Karmouch (2014), the authors proposed a periodical auction-based planning of embedding process in order to handle the incoming set of VN requests. The propounded strategy modeled VNs as bidders competing to win the access to the SN resources. In other words, each request is a selfish bidder that tries to be embedded within Cloud's backbone network. We notice that the proposed approach focuses on the competition between VNs. However, the authors did not detail how can this approach (i.e. selfish) improve the cumulative turnover of CP. In fact, the VNs requests offering the maximum of earnings would have more chance to be accepted regardless of the total revenue generated within the SN.

In Chang et al. (2012), the authors proposed a new algorithm dealing with a multiple virtual network requests embedding (MVNE). The main objectives are (i) maximizing the CP's revenue and (ii) balancing the usage rate of resources in the SN in order to host more VN requests. MVNE algorithm is based on a Mixed Integer Program. Unfortunately, this approach is not scalable. Accordingly, this proposal cannot be exploited in a large scale SN.

In Chowdhury et al. (2012), the authors proposed a new algorithm named Window-based virtual Network Embedding (WiNE) in order to gauge the lookahead impact. It discretizes the time into sequential time-windows and stores the incoming clients' requests to be processed in a batch manner at the end of a time-window. It is worth pointing out that each request is enhanced by a new information which is the maximum waiting period. This parameter is important to specify the maximum duration that a request can be postponed in order to be processed. At the end of this period, if the request is not mapped it would be rejected. The main advantage of WiNE is that it does not depend on the underlying online strategy. WiNE processes as follow. It sorts the stored VNs basing on their revenue. Then it embeds the top of the queue basing on the underlying online strategy. If the mapping of the current request fails due to physical resources lack, so it will be postponed to the next time-window processing. The authors made use of a 50-routers substrate network and varied the arrival rate from 4 to 8 time units per 100 time units. We notice that these parameters are not defined for a scalable scenario and should be stressed more in order to deeply evaluate the performance of WiNE. Besides, sorting the requests basing on revenue could lead to a nonoptimized batch mapping.

4.2 Overview of Virtual Network Embedding with Reliability Constraint

Hereafter, we provide a deep survey of the survivable VN mapping approaches. A taxonomy of this kind of strategies can be defined based on the main following criteria: (i) centralized or distributed, (ii) proactive or reactive, (iii) the type of the protected resource (i.e. router and/or link, geographic region), (iv) backup use, and (v) curative or preventive. It is noteworthy that all reliable VN embedding strategies found in the literature are online approaches.

4.2.1 Distributed Approaches

The authors introduced in Houidi et al. (2010) a new distributed fault-tolerant embedding (DFTE) algorithm. The new propounded strategy deals with router and link failures (or service degradation). Indeed, the proposed strategy relies on a multi-agent system to tackle the survivable VN mapping. In other terms, this new framework is composed of autonomous agents integrated into the substrate routers that carry out a decentralized fault-tolerant VN embedding algorithm to cope with router and link failures. The new approach is basically a two-step strategy: (i) mapping the virtual infrastructure (VInf) request in a distributed manner and (ii) continuously, checks the router and link state (i.e. by sending alive-message) and then reacts when failure is detected. The decentralized embedding algorithm relies on the approach detailed in Houidi et al. (2009). Hereafter, we provide a summary:

- *Resource description and advertisement*: Infrastructure Providers (InPs) describe and customize their offered substrate resources.
- *Resource discovery and matching*: it consists of searching and finding resource candidates that comply with the requirements specified by the $\mathcal{VN}$ request.
- *$\mathcal{VN}$ embedding*: InP finds the optimal $\mathcal{VN}$ embedding in distributed way basing on the earlier steps.
- *$\mathcal{VN}$ binding*: The selected substrate resources are allocated by the InPs in order to instantiate the requested $\mathcal{VN}$.

Once the $\mathcal{VN}$ request is mapped, the proposed framework ensures (i) detecting and identifying local changes through monitoring (e.g. node/link failure, performance degradation), (ii) selecting new substrate resources to maintain $\mathcal{VN}$ topologies operational, (iii) instantiating a virtual router in the new selected substrate router, and (iv) binding the virtual router along with the virtual links that are affected. The framework relies on the multi-agent based approach to ensure distributed negotiation and synchronization between the substrate routers. The negotiation and synchronization are optimized thanks to a clustering based technique. Hereby, we summarize the DFTE algorithm:

- Upon detecting a virtual node failure, send a notification message to all agents in the same cluster.
- Compute dissimilarity metric between the affected router and neighbor routers. It is noteworthy that dissimilarity metric describes the pertinence of the evaluated router. The router having the maximum metric is the worst choice that can be made.
- Exchange, via messages, the computed dissimilarity metrics within the same cluster. The agent compares its dissimilarity metric with all substrate nodes. The physical router having the minimum dissimilarity with the affected one will be selected.
- Map the associated virtual links to the substrate paths using a distributed shortest-path algorithm (Houidi et al., 2008).
- Once link failure is detected, the algorithm proceeds directly to the previous step.

In order to evaluate the performance of the proposed framework, the authors resort to simulations. They focus specially on performance and scalability evaluation. The performance metric is basically the delay incurred by the proposed adaptive $\mathcal{VN}$ algorithm. Substrate topology is composed of ten nodes and $\mathcal{VN}$ request is five-routers. The time required to adapt the $\mathcal{VN}$ with the distributed embedding algorithm is always less than 2 seconds and is decreased further in the presence of multiple clusters. The authors deal with scalability evaluation through GRID-5000 (Grid'5000). It has been used to generate full mesh substrate topologies with different sizes (from 0 up to 100 nodes). Results show that the number of messages exchanged between substrate nodes decreases with clustering. Besides, the proposed strategy keeps a good recovery time from router failure.

It should be highlighted that the authors in this article define their proposal basing on multi-agent system, so all the underlying physical network resources should be updated to include an agent-based mechanism. This is not an easy task to be supported by the InP in the near future. Furthermore, the authors did not explain whether each router in one cluster should have a snapshot of all the routers within its cluster or not. In fact, it is a mandatory task to ensure the migration upon detecting a router failure.

However, this synchronization may overload the network as well as occupy router memory.

We notice, also, that the authors do not evaluate the resource consumption of the network capacities. Moreover, they do not deal with rejection rate of VNs which is directly impacted by recovery mechanism since more physical resources will be used. In fact, this metric expresses the earned benefit.

4.2.2 Centralized Approaches

The main related research work tackling the survivable $\mathcal{VN}$ embedding is addressed based on centralized architecture. In fact, they assume the existence of a centralized controller that not only manages the mapping process but also ensures some level of fault-tolerant mechanism. The centralized strategies can be classified also basing on the protected physical resource(s): (i) router, (ii) link, (iii) router and link, and (iv) geographic zone. Hereby, we summarize them respecting the aforementioned taxonomy.

4.2.2.1 Substrate Router Failures

In Yeow and Kozat (2010), the authors addressed the $\mathcal{VN}$ reliable embedding problem and proposed a new opportunistic redundancy pooling (ORP) algorithm. The main idea behind the proposal is to share redundant (i.e. backup) resources between many $\mathcal{VN}$s. In other words, backups of some $\mathcal{VN}$s are pooled together. In such a way, the number of redundant resources is minimized. The number of backup routers is calculated analytically so that each $\mathcal{VN}$ can guarantee the required level of reliability. ORP operates as following. First, the $\mathcal{VN}$ request is increased with backup links and routers. Then, the problem is formulated as an MILP solved by the open-source coin-or branch and cut (CBC) solver. The aforementioned steps are summarized in the pseudo-code Algorithm 4.1. To evaluate the performance of ORP, it is compared with related $\mathcal{VN}$ reliable embedding strategies assuming no-share of redundancy resources between $\mathcal{VN}$s. Besides, ORP is compared with the baseline approach (i.e. does not consider survivability) in order to gauge the additional amount of resources consumed for reliability. The simulation results show that ORP is better than no-share approaches in terms of rejection rate of $\mathcal{VN}$s and number of required backup resources. In Koslovski et al. (2010), more technical details of ORP are exposed. In fact, the authors extended the virtual execution description language (VXDL) (Koslovski et al., 2008) to enable the specification of reliable VInfs. Unfortunately, ORP does not deal with substrate link failures. Moreover, the match between a critical router and its associated backups is not detailed.

Algorithm 4.1: ORP. Source: (Yeow and Kozat, 2010, Koslovski et al., 2010).

---

1 Inputs: Virtual and substrate networks described by graphs
2 Output: Working and backup mappings
3 Analytically computing the backup routers
4 $\mathcal{VN}$ graph is increased with backup routers and links
5 Mapping problem formulation basing on MILP
6 Solve the relaxed MILP mapping problem using CBC solver

In Yu et al. (2011), the authors tackled the resilience $\mathcal{VN}$ embedding problem by considering substrate routers failures. A new heuristic approach is proposed named survivable virtual infrastructure mapping (SVIM). The latter is based on the K-Redundant scheme for surviving facility router failure ($K \geq 1$), thus it is denoted K-Redundant algorithm. Note a K-Redundant scheme means that K substrate routers are dedicated to serve as a backup for all critical routers. The objectives are (i) to minimize the usage rate of physical resource and (ii) to maximize the reliability of $\mathcal{VN}$s. This optimization problem is formulated as an MILP. K-Redundant solves the above problem based on two stages. First, the $\mathcal{VN}$ request is increased with redundant virtual routers and links attached to the selected routers. Note that this stage processes only the critical virtual routers. In the second stage, the increased $\mathcal{VN}$ request is mapped in the $\mathcal{SN}$ with respect to the usage rate of resource. In order to minimize the total cost of the mapping, K-Redundant makes use of the backup share approach. Indeed, since the authors assumed that only one substrate router can fail at a time, backup paths belonging to different critical virtual routers can share the same bandwidth. Besides, the cross-share approach is adopted. It consists in sharing the original primary paths with the corresponding backup paths. The $\mathcal{VN}$ embedding process is based on D-ViNE (Chowdhury et al., 2012) algorithm to find the working mapping for the original $\mathcal{VN}$ request. Then, K-Redundant solves the simplified MILP using CPLEX to embed the backup resources. These steps are presented in the pseudo-code Algorithm 4.2. The results show that if $K > 1$, the reliability increases whereas the router cost's ratio is higher than 1. Unfortunately, physical links are assumed to remain operational at all times, which is not realistic. Besides, the rejection rate of $\mathcal{VN}$ requests is not evaluated.

4.2.2.2 Substrate Link Failures

In Rahman and Boutaba (2012), the authors proposed a new survivable virtual network embedding algorithm denoted by SVNE. In fact, when

---

Algorithm 4.2: K-Redundant. Source: Modified from Yu et al. (2011).

---

1 Inputs: Virtual and substrate networks described by graphs, k value
2 Output: Working and backup mappings
3 Get the working mapping using D-VINE (Chowdhury et al., 2012)
4 Establish an augmented reliable graph with backup node(s) basing on k value
5 Solve the reduced MILP mapping problem (backup) using CPLEX

---

a substrate link failure occurs, a fast re-routing strategy is executed by exploiting only the backup bandwidth allocated in all physical links. In other words, in order to protect against single substrate link failure, SVNE dedicates a certain rate of its bandwidth for a backup usage. SVNE heuristic is composed of three stages. First, SVNE proactively computes a set of possible backup detours for each substrate link. Then, SVNE embeds each new $\mathcal{VN}$ request by calling the $\mathcal{VN}$ embedding strategy D-ViNE (Chowdhury et al., 2012). Finally, when a link failure occurs, a reactive backup detour optimization solution is invoked. It reroutes the affected bandwidth along candidate backup detours selected in the first stage. To do so, the authors formulated the problem of the $\mathcal{VN}$ embedding problem and re-mapping of virtual links as linear programs. The main objective is to minimize the bandwidth consumption and the penalties due to link failures. The SVNE algorithm is summarized in the pseudo-code Algorithm 4.3. Based on simulations, SVNE outperforms the baseline strategies (i.e. reliability is not considered). Unfortunately, the successful recovery rate of virtual links impacted by failures is not evaluated. Besides, the substrate router failures are not considered.

Algorithm 4.3: SVNE. Source: Modified from Rahman and Boutaba (2012).

---

1 Compute detour paths for each substrate link
2 Modelize the mapping problem on MILP
3 Embed the $\mathcal{VN}$ basing on MILP resolution
4 Re-embed the impacted link using the pre-computed detours and basing on MILP resolution

---

In Yan et al. (2011), the authors propounded RMap algorithm dealing with the mapping of $\mathcal{VN}$s by considering failures of substrate links. To maximize the resilience, RMap allocates backup links. To do so, first RMap embeds the $\mathcal{VN}$ request by embedding virtual routers then deals with virtual links.

Next, the $\mathcal{SN}$ is formulated as a weighted graph by defining for each link its stress value. Note that the latter quantifies the number of virtual links transiting through the substrate link. Afterward, if a link stress is higher than a predefined threshold, RMap will compute its backup detour based on Loop Free Alternate resilience approach (Atlas and Zinin, 2008). When a link failure occurs, virtual links transiting over this substrate link will migrate to the backup links. Hence, the offered service will not be interrupted. Based on simulations, RMap achieves 70% of virtual link protection. However, RMap ensures protection only for stressed links. Thus, if a failure occurs within a nonstressed link, all $\mathcal{VN}$s will be impacted. Furthermore, the rejection rate of $\mathcal{VN}$ requests is not evaluated. Finally, we notice that the calibration of the links' stress threshold has not been discussed.

4.2.2.3 Substrate Router and Link Failures

In Liu et al. (2009), the authors addressed the problem of $\mathcal{VN}$ embedding in the $\mathcal{SN}$. The authors assume that both substrate links and routers failures can occur in the $\mathcal{SN}$. Two strategies dealing with reliability of $\mathcal{VN}$s named (i) cluster and path protection (CPP) and (ii) virtual network protection (VNP) are proposed. With CPP, each logical connection (i.e. virtual link) is protected against substrate link failures by establishing two disjoint paths. Besides, two copies of the job (virtual router) are allocated to survive any single router (i.e. cluster) failure. With VNP, three disjoint paths and three copies of jobs are respectively allocated to survive against substrate links and routers failures. The task graph of a $\mathcal{VN}$ contains m connections and n tasks. Hence, to embed a VN, (i) CPP needs a total of $2 \cdot n$ routers and $2 \cdot m$ pairs of disjoint paths, in contrast (ii) $3 \cdot n$ virtual routers and $3 \cdot m$ connections are necessary with VNP.

CPP operates as following. First, for each unassigned virtual router, the compatible substrate router with minimum cost is selected. Then, virtual connections (primary and backup links) are mapped while maximizing the rate of sharing resources in order to minimize the mapping cost.

On the other hand, VNP computes for each $\mathcal{VN}$ three disjoint mappings. In fact, the embedding algorithm is similar to CPP but without considering backup links. VNP first instantiates the primary mapping. Then, all the physical resources (routers and links) used in the first step mapping are removed to deal with the second mapping (i.e. first backup). Afterward, VNP computes the embedding of the second backup (i.e. third mapping), after removing all the resources used in the first and the second mapping for the same request. Consequently, VNP generates three disjoint $\mathcal{VN}$ mappings.

To evaluate the performance, two metrics are defined: (i) $\mathcal{VN}$ blocking rate and (ii) the average resource leasing cost. Simulation results show that

VNP has a lower $\mathcal{VN}$ blocking rate than CPP when a sufficient computing resources and varying link capacities are assumed. Whereas, CPP has a lower $\mathcal{VN}$ blocking rate than VNP when a sufficient bandwidth and varying router capacities are assumed. Besides, the simulations illustrate that VNP achieves a lower average of job cost than CPP. Unfortunately, both CPP and VNP make use of a huge amount of network resources, which will deteriorate the rejection rate of $\mathcal{VN}$ requests and $C\mathcal{P}$'s benefit. Furthermore, we notice that simulations consider only small-sized $\mathcal{VN}$ requests (three routers), which is not realistic.

In Soualah et al. (2014), Soualah et al. (2017), we proposed a novel centralized (software defined network [SDN] based) reactive survivable network embedding strategy. Our proposal is based on Game theory. Based on extensive network simulations, we illustrate the good performance obtained compared to the above related strategies in terms of quality of service (QoS) satisfaction and provider's revenue.

4.2.2.4 Regional Failures

In Yu et al. (2010b), two survivable $\mathcal{VN}$ embedding algorithms have been proposed named separate optimization with unconstrained mapping (SOUM) and incremental optimization with constrained mapping (IOCM). The challenging point in Yu et al. (2010b) is the consideration of geographic region failures that can simultaneously affect a set of substrate resources (i.e. links and/or nodes). In general, a failure of a geographic region infers a simultaneous outage of nodes and links due to events such as natural disasters, etc. To do so, similar to Rahman and Boutaba (2012), the problem is formulated as an MILP. The proposal embeds the $\mathcal{VN}$ request with respect to the survivability against any single regional failure as follows. First, the $\mathcal{VN}$ request is mapped depending on the adopted algorithm SOUM or IOCM. Then, the redundant nodes and links are allocated. Next, if the regional failure occurs then the virtual resources migrate to the backup resources. SOUM calculates the mapping of $\mathcal{VN}$ request regardless of any regional failures. Afterward, for each regional failure it instantiates the backup resources in safe region(s). It is worth pointing out that the order of regional failures does not affect the mapping result. Unlike IOCM, the mapping depends on the order of dealing regional failures. The proposal has been evaluated and a comparison is performed between the SOUM and IOCM in terms of (i) mapping cost, (ii) average number of migrations, and (iii) recovery blocking probability. Unfortunately, the rejection rate of $\mathcal{VN}$s is not evaluated. Moreover, the authors did not compare the new approaches with baseline solutions (i.e. algorithms that do not consider reliability) basing on reject rate metric in order to evaluate the impact of allocating dedicated backup. Besides, the two proposed algorithms consume a huge amount of network capabilities to ensure survivability without an optimized strategy.

In Yu et al. (2010a), authors extend their work in Yu et al. (2010b) by first developing a non-survivable virtual infrastructure mapping (NSVIM*) heuristic. Based on NSVIM*, they develop efficient SVIM heuristics namely separate optimization with unconstrained mapping and redundancy elimination (SOUM*) and incremental optimization with constrained mapping (IOCM*) and failure-avoidance. In addition, they also develop an MILP formulation to model the SVIM problem with the objective of minimizing the overall cost. In this paper, the authors assume that two (or more) regional failures cannot simultaneously occur. Like Yu et al. (2010b), the authors ensure virtual infrastructure (VI), also called virtual network $\mathcal{VN}$, survivability against regional failures, so the inputs for SOUM* and IOCM* are (i) the substrate network, (ii) $\mathcal{VN}$ request, (iii) list of possible regional failures R, and (iv) the incoming requests. Similar to the NSVIM algorithm (Yu et al., 2010b), the improved-NSVIM (NSVIM*) approach, developed in this paper, satisfies the $\mathcal{VN}$ requests without any survivability requirement. Compared to NSVIM (Yu et al., 2010b), NSVIM* uses an efficient $\mathcal{VN}$ node sorting strategy to map those $\mathcal{VN}$ nodes first that require large computing and bandwidth resources. Hereafter, we summarize the main stages of NSVIM*:

- Sort the virtual routers by their degree.
- Choose a router with the highest degree.
- Find the set of substrate routers that can host the selected virtual router, if this set is empty the virtual request is rejected.
- Choose the substrate router with low cost.
- Ensure the connection between this router and its neighbor(s) which is (are) already mapped.
- Iterate this procedure (i.e. the aforementioned steps) until all virtual routers are mapped.

Similar to SOUM, SOUM* decomposes the SVIM problem into $|R| + 1$ separate NSVIM problems: (i) one involving the initial pre-failure (i.e. working) mapping and (ii) the others involving the after-failure (i.e. backup) mapping. The challenging idea with SOUM* comparing to SOUM is that SOUM* eliminates the redundancy mappings that cover the same regional failure by keeping the min-cost mapping. Let us take the example of two mappings $M1$ and $M2$. The first covers two regions $R1$ and $R2$. But $M2$ covers only $R1$. Besides, $M1$ cost is lower than $M2$ one. Accordingly, SOUM* keeps $M1$ and eliminates $M2$. At the end of SOUM* algorithm, the authors should have a min-cost set of mappings that covers whole regional failures. SOUM* can be summarized as below:

- Calculate the mapping for a specific regional failure. It also includes initial pre-failure (working) mapping.
- Calculate the cost of this mapping.
- Remove the redundant mapping(s) by keeping the low cost mapping.
- Iterate this procedure to all regional failures.

On the other hand, an optimized version of IOCM (Yu et al., 2010b) denoted by IOCM* is proposed. The latter aims to minimize the additional computing and networking resources required to recover from physical failures. Similar to IOCM, the order in IOCM* impacts the whole mapping cost. IOCM* is based on new version of NSVIM* called failed avoidance NSVIM* (FA-NSVIM*). It tries to avoid the mapping in a facility router within a regional failure. Hence, the backup number and the whole mapping cost will be reduced. Hereinafter, we summarize IOCM*.

1. Consider k order of $|R|$ regional failures and pre-failure mapping (i.e. $k < (|R| + 1)!$).
2. Call FA-NSVIM* algorithm and get the initial mapping, which is affected the least by failures.
3. Basing on NSVIM*, remap all substrate resources (i.e. links and routers) which are in the same specific regional failure, into other physical resources.
4. Iterate the previous step for all regional failures.
5. Calculate the cost of the current mapping (i.e. primary and backup resources).
6. Keep the current mapping if it is lower than its predecessor.
7. Consider another order (i.e. from the k order) and start from the second step.

To evaluate the performance of the two approaches, the authors use (i) small substrate network: 10 routers and 15 links and (ii) large substrate network: 27 routers and 41 links. The computing capacity at facility routers and bandwidth capacity on the links follow a uniform distribution from 300 to 600 under unconstrained capacity and 50 to 150 when the capacity is constrained. They assume that the unit computing cost (e.g. using 1000 CPU hours) is uniformly distributed from 1 to 5 and unit bandwidth cost (e.g. for 1 Mbps) is from 5 to 10. The $\mathcal{VN}$ requests are generated randomly. The computing and bandwidth demands follow a uniform distribution from 10 to 30 and 10 to 50, respectively. For each regional failure, they randomly choose various numbers of adjacent routers to fail (e.g. three routers). The authors used the following three performance metrics: (i) the cost, (ii) the mean number of migrations, and (iii) the recovery blocking probability, which is the ratio of the number of unrecoverable failure scenarios to the total number of failure scenarios. In order to evaluate the cost NSVIM*, the authors compare it to NSVIM and MILP problem solved by CPLEX 8.0 but for a very small $\mathcal{VN}$ request. Results show that NSVIM* and NSVIM have a near-optimal (i.e. same cost as the MILP) performance for a small-size $\mathcal{VN}$ request. Furthermore, NSVIM* achieves better performance than NSVIM when the size of the $\mathcal{VN}$ request increases. Besides, results show that SOUM* and IOCM* achieves a near-minimum cost and perform better than SOUM and IOCM for a small-size problem. For lager $\mathcal{VN}$ request, the proposed algorithms SOUM*, IOCM*, and IOCM have similar performance in terms of cost, and each one of them outperforms SOUM. Moreover, simulation proves the proposed algorithms IOCM* and IOCM lead to fewer average number of migrations than SOUM* and SOUM. Furthermore, results show that IOCM* leads to a fewer migrations than IOCM. This is explained by the use of FA-NSVIM* algorithm. Besides, results show that the SOUM* and SOUM lead to better recovery blocking probability than IOCM* and IOCM. We notice that the authors propounded their approaches basing on the assumption that only one regional failure fails at a time. But, this assumption is not realistic since many failures can simultaneously occur in different geographical regions. Furthermore, authors did not specify the path use between the primary and the backup node. Besides, they did not determine the period and bandwidth allocation for synchronization (i.e. sending snapshots). In fact, this period can cause communication degradation. On the other hand, authors did not evaluate the rejection rate. Indeed, this metric shows the approach efficiency to optimize the allocation of substrate resources.

In Sun et al. (2011), the authors devise two kinds of algorithms for solving the survivable virtual network mapping (SVNM) problem efficiently: (i) Lagrangian relaxation-based (LR-SVNM) algorithms including LR-SVNM-M and LR-SVNM-D and (ii) Heuristic (H-SVNM) algorithms including H-SVNM-D and H-SVNM-M. Similar to Yu et al. (2010b), the authors propose survivable solutions to face regional failures: Shared Risk Group (SRG). The authors assume that only one region can fail at a time. These algorithms mainly aim to decompose the primal NP-hard problem into several sub-problems in order to reduce the computational complexity at the expense of increasing the total $\mathcal{VN}$ mapping cost. At the first step, the authors formulate the problem using MILP approach. The MILP formulation is similar to Chowdhury et al. (2009). The first propounded algorithm (i.e. H-SVNM) is an enhancement of SOUM* (Yu et al., 2010a). In fact, the authors propose a modified version of D-ViNE algorithm (Chowdhury et al., 2009), named D-ViNE*. Like SOUM* (Yu et al., 2010a), H-SVIM decompose the initial problem onto sub-problems basing on the number of regional failures, then it uses D-ViNE* to deal with each mapping. This approach is called H-SVIM-D. Alternatively, the embedding of each sub-problem can be done by solving the MILP formulation if the problem is tractable. This algorithm is called H-SVIM-M. Like SOUM*, once getting the mapping of all sub-problems, H-SVIM-M or H-SVIM-D use the greedy min-cost set cover algorithm (Yu et al., 2010a) to eliminate redundancies. On the other hand, the authors define new approach basing on Lagrangian relaxation method. This new proposal is named LR-SVNM. The main idea is to decompose the NP-hard problem into many sub-problems by relaxing certain constraints. Each one of these sub-problems may be solved using D-ViNE* (i.e. LR-SVNM-D), or by MILP (i.e. LR-SVNM-M). To evaluate the performance of the propounded algorithms, the authors used four substrate network topologies in their simulations. All router and link bandwidth capacities are assumed to be 50 units. VN requests are generated randomly. The number of $\mathcal{VN}$ routers is equal to a given number N and the average degree of connectivity of the $\mathcal{VN}$ request is about 2.5. They assume that each $\mathcal{VN}$ router requires 1 unit computing resource capacity and 1 unit of bandwidth resources by each of the communication demands between the $\mathcal{VN}$ nodes. MILP problem is solved by the CPLEX solver. The authors compared IOCM*, SOUM*, LR-SVNM-D, LR-SVNM-M, H-SVNM-D, H-SVNM-M, and MILP. The performance metrics are (i) total mapping cost and (ii) time efficiency which is the time consumed by the algorithm to map a $\mathcal{VN}$. The simulation results show that the proposed algorithms have mapping cost lower than IOCM* and SOUM*. In small topologies, curves show that LR-SVNM-M and H-SVNM-M outperform LR-SVNM-D and H-SVNM-D, respectively. Besides, H-SVNM-M is better than LR-SVNM-M in term of mapping cost. On the other hand, IOCM* and SOUM* are considerably better than the proposed algorithms in term of time efficiency. Moreover, H-SVNM is better than the Lagrangian relaxation-based algorithm LR-SVNM in terms of computational time. We notice that the authors propound their approaches basing on the assumption that only one regional failure fails at a time. But, this assumption is not realistic since many failures can occur in different geographical regions simultaneously (e.g. earthquake). We notice, also, that the authors did not detail the synchronization procedure. Furthermore, the authors use a modified version of D-ViNE (Chowdhury et al., 2009), but this algorithm has two parameters constraining the mapping process which are location and distance. Moreover, simulation results show that

Table 4.1 Overview of reliable $\mathcal{VN}$ embedding strategies

Algorithm	Strategy	Protection	Link mapping	Backup	Recovery	Preventive	Proactive
ORP (Yeow and Kozat, 2010, Koslovski et al., 2010)	Centralized	Router	Splittable	Yes	Yes	No	Yes
K-Redun-dant (Yu et al., 2011)	Centralized	Router	Splittable	Yes	Yes	No	Yes
RMap (Yan et al., 2011)	Centralized	Link	Unsplittable	Yes	Yes	No	Yes
SVNE (Rahman and Boutaba, 2012)	Centralized	Link	Splittable	Yes	Yes	No	Yes
CPP, VNP (Liu et al., 2009)	Centralized	Router and link	Unsplittable	Yes	Yes	No	Yes
SOUM, IOCM (Yu et al., 2010b)	Centralized	Regional	Unsplittable	Yes	Yes	No	Yes
SOUM* (Yu et al., 2010a)	Centralized	Regional	Unsplittable	Yes	Yes	No	Yes
IOCM* (Yu et al., 2010a)	Centralized	Regional	Unsplittable	Yes	Yes	Yes	Yes
LR-SVNM-M, LR-SVNM-D (Sun et al., 2011)	Centralized	Regional	Splittable	Yes	Yes	No	Yes
H-SVNM-D, H-SVNM-M (Sun et al., 2011)	Centralized	Regional	Splittable	Yes	Yes	No	Yes
DFTE (Houidi et al., 2010)	Distributed	Router and link	Splittable	No	Yes	No	No

`LR-SVNM-M` and `H-SVNM-M` spend so much time to compute the $\mathcal{VN}$ embedding for networks having more than 10 routers. Accordingly, they are not scalable.

4.2.3 Summary

Table 4.1 summarizes the main survivable VN embedding algorithms found in the existing literature. Unfortunately, all of them are online algorithms and to the best of our knowledge no batch strategy has been proposed to deal with the reliable embedding of $\mathcal{VN}$ requests.

4.3 Conclusion

This chapter summarized the most prominent VN mapping strategies interconnecting data centers. First, we addressed the strategies which do not consider any fault-tolerant mechanism. Later on, we presented an overview of the survivable VN embedding algorithms found in the literature. It is worth pointing out that all the proposed approaches sequentially process the $\mathcal{VN}$ requests as soon as they arrive.

References

A. Atlas and A. Zinin. Basic specification for IP fast reroute: loop-free alternates. *RFC 5286*, 2008.

X. Chang, B. Wang, and J. Liu. Network state aware virtual network parallel embedding. *IEEE Performance Computing and Communications Conference (IPCCC)*, pages 193–194, 2012.

N.M.M.K. Chowdhury, M.R. Rahman, and R. Boutaba. Virtual network embedding with coordinated node and link mapping. *IEEE INFOCOM*, pages 783–791, 2009.

M. Chowdhury, M. Rahman, and R. Boutaba. ViNEYard: virtual network embedding algorithms with coordinated node and link mapping. *IEEE/ACM Transactions on Networking (TON)*, 20: 206–219, 2012.

M. Dorigo and C. Blum. Ant colony optimization theory: a survey. *Theoretical Computer Science*, 344: 243–278, 2005. ISSN 0304-3975. doi: 10.1016/j.tcs.2005.05.020. URL http://www.sciencedirect.com/science/article/pii/S0304397505003798.

I. Fajjari, N. Aitsaadi, G. Pujolle, and H. Zimmermann. VNE-AC: virtual network embedding algorithm based on ant colony metaheuristic. *IEEE ICC*, 2011a.

I. Fajjari, N. Aitsaadi, G. Pujolle, and H. Zimmermann. VNR algorithm: a greedy approach for virtual networks reconfigurations. pages 1–6, 2011b.

I. Fajjari, N. Aitsaadi, G. Pujolle, and H. Zimmermann. Adaptive-VNE: a flexible resource allocation for virtual network embedding algorithm. *IEEE GLOBECOM*, pages 2640–2646, 2012.

I. Fajjari, N. Aitsaadi, M. Pioro, and G. Pujolle. A new virtual network static embedding strategy within the cloud's private backbone network. *Elsevier Computer Networks*, 2014. 62. 69–88.

Grid'5000. GRID-5000 Platform. https://www.grid5000.fr/.

I. Houidi, W. Louati, and D. Zeghlache. A distributed virtual network mapping algorithm. *ICC*, pages 565–578, 2008.

I. Houidi, W. Louati, D. Zeghlache, and S. Baucke. Virtual resource description and clustering for virtual network discovery. *IEEE ICC Workshop*, 2009.

I. Houidi, W. Louati, D. Zeghlache, P. Papadimitriou, and L. Mathy. Adaptive virtual network provisioning. *ACM SIGCOMM Workshop on Virtualized Infrastructure Systems and Architectures (VISA)*, pages 41–48, 2010.

A. Jarray and A. Karmouch. Decomposition approaches for virtual network embedding with one-shot node and link mapping. *IEEE/ACM Transactions on Networking*, 2014. 23. 1012–1025

J. Kleinberg. Approximation algorithms for disjoint paths problems. *PhD thesis*, MIT, 1996.

S.G. Kolliopoulos and C. Stein. Improved approximation algorithms for unsplittable flow problems. *Proceedings of the IEEE Symposium on Foundations of Computer Science*, 1997.

G. Koslovski, P.V.-B. Primet, and A.S. Charao. VXDL: virtual resources and interconnection networks description language. *GridNets*, 2008.

G. Koslovski, W.-L. Yeow, C. Westphal, T.T. Huu, J. Montagnat, and P. Vicat-Blanc. Reliability support in virtual infrastructures. *International Conference on Cloud Computing Technology and Science (CLOUDCOM)*, pages 49–58, 2010.

X. Liu, C. Qiao, and T. Wang. Robust application specific and agile private (ASAP) networks withstanding multi-layer failures. *IEEE Optical Fiber Communication*, 2009.

M. Rahman and R. Boutaba. SVNE: survivable virtual network embedding algorithms for network virtualization. *IEEE Transactions on Network and Service Management (TNSM)*, pages 1–14, 2012.

O. Soualah, I. Fajjari, N. Aitsaadi, and A. Mellouk. A reliable virtual network embedding algorithm based on game theory within cloud's backbone. *IEEE ICC*, 2014.

O. Soualah, N. Aitsaadi, and I. Fajjari. A novel reactive survivable virtual network embedding scheme based on game theory. *IEEE Transactions on Network and Service Management*, 2017. 14. 569–585.

T. Stützle and H.H. Hoos. MAX-MIN ant system. *Future Generation Computer Systems*, 16: 889–914, 2000.

G. Sun, H. Yu, L. Li, V. Anand, and H. Di. The framework and algorithms for the survivable mapping of virtual network onto a substrate network. *IETE Technical Review*, 2011. 28. 381–391.

C. Wang, Y. Yuan, C. Wang, X. Hu, and C. Zheng. Virtual bandwidth allocation game in data centers. *IEEE International Conference on Information Science and Technology*, 2012.

W. Yan, S. Chen, Xin Li, and Yan Wang. RMap: an algorithm of virtual network resilience mapping. *International Conference on Wireless Communications, Networking and Mobile Computing (WiCOM)*, pages 1–4, 2011.

W.-L. Yeow, C. Westphaland, and U.C. Kozat. Designing and embedding reliable virtual infrastructures. *ACM SIGCOMM Workshop on Virtualized Infrastructure Systems and Architectures (VISA)*, pages 33–40, 2010.

M. Yu, Y. Yi, J. Rexford, and M. Chiang. Rethinking virtual network embedding: substrate support for path splitting and migration. *SIGCOMM Computer Communication Review*, 38: 17–29, 2008.

H. Yu, C. Qiao, A. Vishal, and X. Liu. On the survivable virtual infrastructure mapping problem. *IEEE ICCCN*, 2010a.

H. Yu, C. Qiao, V. Anand, X. Liu, H. Di, and G. Sun. Survivable virtual infrastructure mapping in a federated computing and networking system under single regional failures. *IEEE Global Telecommunications Conference (GLOBECOM)*, pages 1–6, 2010b.

H. Yu, V. Anand, C. Qiao, and G. Sun. Cost efficient design of survivable virtual infrastructure to recover from facility node failures. *IEEE International Conference on Communications (ICC)*, pages 1–6, 2011.

Y. Zhou, Y. Li, G. Sun, D. Jin, L. Su, and L. Zeng. Game theory based bandwidth allocation scheme for network virtualization. *IEEE Global Communications Conference*, 2010.

5

An Evaluation Method of Optimal Cost Saving in a Data Center with Proactive Management

Ruben Milocco[1], Pascale Minet[2], Éric Renault[3], and Selma Boumerdassi[4]

[1] *GCAyS, UNComahue, Neuquén, Argentina*
[2] *Inria, Paris, France*
[3] *LIGM, Univ. Gustave Eiffel, CNRS, ESIEE Paris, Marne-la-Vallée, France*
[4] *CEDRIC, CNAM, Paris, France*

Abstract

The energy consumption of data centers (DCs) has always been increasing. It will exceed 1200 TWh in 2020. In this chapter, we determine whether resource allocation in DCs can satisfy the following three requirements: (i) give short response times, (ii) keep the data center efficient, and (iii) reduce the carbon footprint.

An efficient way to reduce the energy consumption in a DC is turning off servers that are not used for a minimum duration. The high dynamicity of the jobs submitted to the DC requires it to periodically adjust the number of active servers to meet job requests. This is called dynamic capacity provisioning. This provisioning can be based on prediction. In such a case, a proactive management of the DC is performed. The goal of this chapter is to provide a methodology to evaluate the energy cost reduction brought by proactive management, while keeping a high level of user satisfaction.

This chapter presents a method to compute the upper bound of relative cost savings obtained by proactive management with regard to a pure reactive management based on the last value. With this method, it becomes possible to quantitatively compare the efficiency of different predictors.

We also show how to apply this method to a real DC and how to select the value of the DC parameters to get the maximum cost savings. Two types of predictors are evaluated: linear predictors represented by the autoregressive-moving-average (ARMA) model and nonlinear predictors obtained by maximizing the conditional probability of the next sample, given the past. Both are applied to a real data set collected over one month to evaluate the cost savings achieved with these two predictors that are adjusted to maximize energy cost saving.

After reading this chapter, the reader should be able to:

- Understand the principles of reactive and proactive strategies.
- Evaluate an energy cost based on the energy consumption and the balance between the requested and the provided energy.
- Select the predictor minimizing the energy cost among any class of predictors.
- Compute the number of servers to turn on/off according to the predicted energy consumption.
- Apply the methodology proposed to a real DC.

5.1 Introduction

The Network World journal (https://www.networkworld.com/article/3324 050/10-predictions-for-the-data-center-and-the-cloud-in-2019.html) observes that due to the increasing demand in processing power, data center (DC) growth will continue. In addition, machine learning techniques will play a major role in DC management, by optimizing the DC resources through continuous monitoring and adjustment.

The state-of-the art shows that appropriate management optimizes the operational cost of the DC, either by improving the quality of service (QoS) or by saving energy. An efficient way to reduce the energy consumption in a DC is turning off servers that are not used for a minimum duration. The high dynamicity of the jobs submitted to the DC requires it to periodically adjust the number of active servers to serve job requests. This is called dynamic capacity provisioning (DCP). This provisioning can be based on prediction. In such a case, a proactive management of the DC is performed. As a consequence, there is a great interest in studying different proactive strategies based on the prediction of either the energy or the resources needed to serve CPU and memory requests. The cost depends on (i) the proactive strategy used, (ii) the workload requested by jobs, and (iii) the prediction used. The problem complexity explains why, despite its importance, the maximum cost saving has not been evaluated in theoretical studies.

Figure 5.1 Proactive u_p and reactive u_r actions to provide requested energy y.

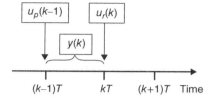

From the energy point-of-view, the DC management can be summarized as follows. The energy requested to serve jobs can be provided either proactively (i.e. a priori) or reactively (i.e. a posteriori), or as a combination of both. In the following, the latter is considered. In Figure 5.1, the scheme of proactive (u_p) and reactive (u_r) actions is depicted.

The purpose of this chapter is to analyze a proactive DC management strategy that minimizes a proposed cost when serving the jobs requesting CPU and memory. A proactive DC management consists in using prediction to configure the DC servers a priori, whereas a reactive DC management consists in configuring servers a posteriori in order to serve the resource requests.

If the prediction matches the energy needed to serve the requests in the current DC configuration period, these requests are served. If the prediction is smaller than the energy needed to serve the requests, there will not be enough switched-on servers in the DC to serve all the arriving requests. The requests that cannot be served with the available resources in the current period have to wait for the next DC configuration period T, when some servers may be added (i.e. reactive action). If the prediction is greater than the requested energy, the DC resources will be under-utilized and a part of the consumed energy will be wasted. Therefore, the objective is to find optimal predictors that minimize the energy cost defined in Section 5.4.

The methodology proposed in this chapter lies in the direct prediction of the energy consumed in the next time interval, based on the recent past, whereas classical approaches predict resource requests instead of energy requests. It also considers that the impact of overestimation and underestimation are not the same. Hence, minimizing the absolute value of the prediction error does not necessarily maximize the energy cost saving.

The remainder of this chapter is organized as follows: Section 5.2 presents a brief state-of-the-art about DC management based on prediction. Section 5.3 defines our framework for modeling the energy consumption in a DC. Section 5.4 proposes a methodology and an energy cost formulation to evaluate the upper bound of the relative energy cost saving that can be achieved by a predictive DC management. Section 5.5 deals with practical issues and presents the results obtained on a real DC. Finally, Section 5.6 concludes the chapter.

5.2 Related Work

To model DC energy consumption, Dayarathna et al. (2016) studied 200 models classified into two hierarchical classes. The software-centric class deals with applications (e.g. MapReduce), OS, virtualization, and machine learning, whereas the hardware-centric class distinguishes between the power conditioning system, the cooling system, optical networks, network devices, and servers. A server includes server storage, memory, processors (single/multi-core, GPU or not). The most frequent energy model for servers is the linear model, where the energy consumed by the server varies linearly with its workload (Chen et al., 2013, Xiao et al., 2013). This model is also adopted in this chapter.

Different solutions have been proposed to improve energy efficiency in cloud computation. For instance, Wolke et al. (2014) showed that for transactional business applications, dynamic resource allocation with live migration does not increase energy efficiency compared to the static allocation of virtual machines to servers. However, many authors Reiss et al. (2012) and Minet et al. (2018) have shown that the DC workload is more heterogeneous and varies more over time. This leads to using dynamic approaches based on prediction in this chapter.

Quasar Delimitrou and Kozyrakis (2014) aims at increasing resource utilization in clusters, while providing high application performance. Users do not express their resource requests but only their performance constraints for each workload type (e.g. latency-critical, Hadoop framework, single node). Quasar uses classification techniques to estimate the application performances when the number of servers, the server type, the amount of resources within a server, and the interference from other workloads vary. The result of this classification allows Quasar to jointly perform resource allocation and assignment.

Zhang et al. (2014) apply the DCP approach to clouds with heterogeneous workloads and heterogeneous servers. DCP aims at reducing energy consumption by adjusting the number of servers on to the workload, while keeping small scheduling delays for user satisfaction. The heterogeneous workload is classified into multiple task classes where each class groups tasks with similar resource requests and performance objectives. More precisely, tasks are first classified according to their CPU and memory requests with k-Means. Since many tasks are short, each task is initially assumed to be short and if not, its duration is labeled long. Zhang et al. formulate DCP as an optimization problem which considers both server and workload heterogeneity, and reconfiguration costs. In this model, time is divided into intervals of equal duration, and control decision about the number of servers to turn on or off is made at the beginning of each time interval, based on the usage of resources (CPU and memory) predicted by autoregressive integrated moving average (ARIMA). The authors show that their online control algorithm based on their model predictive control framework, called Harmony, can reduce energy consumption by 28% when compared to solutions that do not take into account workload and server heterogeneity. In this chapter, a DC management control based on time intervals and aware of servers heterogeneity is adopted.

An energy-efficient resource provisioning framework for cloud data centers is proposed in Dabbagh et al. (2015). The authors define classes of requests by means of a k-Means clustering, where each class groups requests with similar characteristics. In addition, each request is assumed to have resource demands close to the resource demands of the centroid (i.e. the representative) of its class. For each request class, one Wiener filter is used to predict the number of requests that will arrive in the next time interval. These predictions are used to compute the number of servers that must be on to serve these requests. To save energy, servers that are not needed are switched to sleep mode. The authors have applied their framework to the same Google data set used in this chapter. With their solution, 100 MJ/min are saved when compared to the no-power-management scheme. Moreover, 50 MJ/min are saved when compared to a physical-server prediction scheme with some fixed overprovisioning to reduce user dissatisfaction.

This chapter shares with Zhang et al. (2014) and Dabbagh et al. (2015) the same common goal which consists in improving energy efficiency in DCs by using prediction: servers that should be unused according to the prediction are turned off. However, our solution differs from Zhang et al. (2014) and Dabbagh et al. (2015) in the way the goal is to be reached. Both studies, (i) predict the resource requests in the next time interval, (ii) use a placement algorithm to assign requests to servers, and (iii) compute the energy consumed. We directly predict the energy consumed from historical values of energy consumption. We evaluate the relative energy cost saving obtained with regard to a reactive management and give its upper bound.

Google made publicly available a set of traces collected in one of its DCs over a period of 29 days (Reiss et al., 2011). Analyzing these data help researchers to build efficient placement and scheduling algorithms (Beaumont et al., 2014) based on accurate models of servers, jobs, and tasks. For instance, in order to minimize the latency and processing cost of CPU-intensive jobs, Pacini et al. (2018) propose schedulers based on ant colony optimization and particle swarm optimization, used for the selection of both the DC and the physical servers to run the job. Machine learning techniques (e.g. classification, prediction) are widely proposed to improve DC management. For instance, classification has been applied to servers in Minet et al. (2018), where all servers of a same group have the same CPU capacity and the same memory capacity. Classification has also been applied to jobs, as for instance in Alam et al. (2015), where k-Means is used to identify three classes of jobs: short jobs which are frequent and low resource consumers, medium jobs which are less frequent but are great memory consumers, and long jobs which are not frequent but are great CPU consumers. In Yousif and Al-Dulaimy (2017), two task classes are brought to light using k-Means and density-based clustering: CPU-intensive and memory-intensive. A placement algorithm, based on this classification, is proposed. Cao et al. (2018) use random forests to predict the server workload in a DC and automatically detect overload.

Based on time series predictors, Di et al. (2012) showed that the smaller the prediction error the more accurate the management decision. The simplest predictors are the linear ones in which their parameters are estimated from a linear model using a set of observed values. The prediction is computed as a linear combination of the previous observations and possibly their prediction errors. The most famous family of linear predictors is Auto Regressive Moving Average with eXogenous (ARMAX) signal. Liu et al. (2016) show that moving average (MA), auto regressive (AR), and weighted moving average (WMA) predictors give smaller prediction errors with lower complexity compared to neural networks. Prevost et al. (2011) came to the same conclusion: AR is far better than neural networks in predicting the load demand in DCs. ARMA predictors are also used to predict the number of job arrivals in the next 30 minutes, with a mean absolute percentage error (MAPE) close to 0.38 (Yoon et al., 2017).

The nearest-neighbor method is a popular nonlinear approach (Hastie et al., 2009) based on the observations of states and their subsequent evolution. The prediction of a not yet observed state is computed from the most similar states already observed. Since it is unlikely to find the same state, the method finds some closest states and predicts the average of their behavior. In this chapter, the concept of conditional probability is used as in Caires and Ferreira (2005).

As a conclusion, there are many studies dealing with proactive DC management based on user request prediction. However, to the best of our knowledge, there are no studies evaluating the maximum saving in energy cost that can be achieved by proactive management.

5.3 Framework for DC Modeling

5.3.1 Notations and Assumptions

Most DCs have heterogeneous servers from several generations. We classify in a same group all servers with the same CPU and memory capacities, as well as the same instruction execution time and energy consumption parameters. Our approach should be applied to each group of servers. In the following, we focus on any group of servers. The notations used in this chapter are listed in Table 5.1.

For the energy cost saving evaluation, the following assumptions are used:

- Assumptions with regard to jobs: Job requests are represented by a stochastic process with the following assumptions:
  - A_0 Job arrivals are random.
  - A_1 Amounts of requested resources are random.
  - A_2 Request durations are random.
- Assumptions with regard to servers and DC management:
  - A_{10} DC management is done periodically with period T. The time interval $[(k-1)T, kT)$ is also denoted step k for simplicity reasons.
  - A_{11} DC management is in charge of maintaining the number of servers on in each group, proportional to the amount of resources used by this group during the period T. It is the real use or a predicted use. This is done by either turning on the necessary servers or turning off the unnecessary ones.
- Assumptions with regard to energy:
  - A_{20} When a job is waiting to be scheduled, it does not consume any energy.
  - A_{21} CPU and memory resources are allocated for the job execution duration.

5.3.2 Energy Computation

5.3.2.1 Single-Resource Case

For the sake of simplicity, we first focus on the single-resource case. All requests concern the same resource which is either CPU or memory. Any request arriving at time t_i and requiring an amount of resource r_i for a duration τ_i can be represented as a rectangular pulse with amplitude r_i of duration τ_i and zero out of this range, see Figure 5.2 for an illustrative scheme. The area under the pulse represents the energy consumed to serve the request. For example, Figure 5.2 depicts several requests arriving in the interval $[(k-1)T, kT)$.

Table 5.1 Notations

	Name	Meaning
DC mgt	T	DC management period, also called sampling interval
	$r(t_i, r_i, \tau_i)$	resource request arriving at time t_i
		Requesting an amount of resource r_i for a duration τ_i
	$m(t)$	Number of machines on in the group considered at time t
	$r(t)$	Amount of resources used on servers in this group and not yet released at time $t \geq 0$
Consumed energy	$y(k)$	Energy consumed by a group of servers in the time interval $(k-1)T \leq t < kT$
Predicted energy	$\hat{y}(k)$	Prediction of $y(k)$, done at time $(k-1)T$
Prediction error	$e(k)$	Error of prediction done for step k, known at time kT, $e(k) = y(k) - \hat{y}(k)$
Power	P_{idle}	Power consumed at null load by a server
	μ	Coefficient relating the number of servers on to the resources used in the group considered
	η	Power variation coefficient as a function of load for any server in that group
	$P(t)$	Power consumed by the group considered at time t
Energy	c_e	Cost of 1 J paid to the DC energy provider, $c_e > 0$
cost	c_u	Cost of an underestimation of 1 J, $c_u \geq 0$
	c_o	Cost of an overestimation of 1 J, $c_o \geq 0$
	α, β	$\alpha = c_u/c_e$ and $\beta = c_o/c_e$, with α and $\beta \geq 0$
	$J_{\alpha,\beta,T}$	Energy cost using proactive management
	$L_{\alpha,\beta,T}$	Energy cost using only reactive management
	$RES_{\alpha,\beta,T}$	Relative energy cost saving
	$UB_{\alpha,\beta,T}$	Upper bound of $RES_{\alpha,\beta,T}$

Figure 5.2 Requests processed in the time interval $[(k-1)T, kT)$, in gray $r(t)$; $y(k)$ is the energy consumed during the interval T to serve requests; $\hat{y}(k)$ is the energy predicted; and $y(k-1)$ is last energy value.

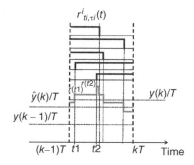

To take server heterogeneity into account, servers are organized into groups. The power consumed by a DC is the sum of the power consumed by each server group. In the following, we only focus on a given group. We express $P(t)$ the power consumed at time $t \in [(k-1)T, kT)$ by a given group as the sum of the power consumed by the servers on belonging to this group. For each server that is on, we adopt the linear power model, which has been extensively used by many authors, Chen et al. (2013) and Xiao et al. (2013), where the power consumed by a server at time t is expressed as the sum of the power consumed at null load and the power proportional to the server load. We get:

$$P(t) = P_{idle}m(t) + \eta\frac{r(t)}{C_{ap}} \tag{5.1}$$

where P_{idle} denotes the power consumption at null load of a server in the server group considered; $m(t)$ is the number of servers in that group that are on at time t; η is its power variation coefficient as a function of the load for any server in that group, C_{ap} is the capacity of the resource considered on any server in that group. It is expressed in absolute (e.g. 1MB of memory); $r(t)$ denotes the amount of resources used at time t on servers in the group considered.

Since the number of servers on in the group considered is proportional to the workload in this group, according to Assumption A_{11}, we can write $m(t) = \mu\frac{r(t)}{Cap}$, where μ is the proportionality coefficient relating the number of servers of the same group to the resource utilization factor. Hence,

$$P(t) = (P_{idle}\mu + \eta)\frac{r(t)}{C_{ap}} \tag{5.2}$$

The energy consumed by a given group to serve the requests in the time interval $[(k-1)T, kT)$ is:

$$y(k) = \int_{(k-1)T}^{kT} P(t)dt = \left(P_{idle}\mu + \eta\right)\int_{(k-1)T}^{kT}\frac{r(t)}{C_{ap}}dt \tag{5.3}$$

5.3.2.2 Extension to the Multi-resource Case

The results obtained so far apply to the case of single-resource requests, such as either CPU or memory. However, extension to the multi-resource case can be done as follows. To compute the power consumed at time $t \in [(k-1)T, kT)$ to serve requests on several resources (e.g. CPU, memory, disk, etc.), we need to determine the number of servers that are on at time t in the group considered. This number of servers on is determined by the bottleneck resource (Zhang et al., 2014). The bottleneck resource for this group in the time interval $[(k-1)T, kT)$, denoted b, is defined as the resource b in the group that maximizes the resource utilization factor $\frac{r_b(t)}{Cap_b}$ in this time interval. The energy consumed in the time interval $[(k-1)T, kT)$ can be written:

$$y(k) = \left(P_{idle}\mu + \eta\right) \int_{(k-1)T}^{kT} \frac{r_b(t)}{Cap_b} dt + \eta \sum_{j \neq b,\ j \in \mathcal{R}} \int_{(k-1)T}^{kT} \frac{r_j(t)}{Cap_j} dt \qquad (5.4)$$

where $r_b(t)$ is the amount of the bottleneck resource b used at time t in the group considered; Cap_b is the capacity of the bottleneck resource b on any server of this group; $\mathcal{R}$ denotes the set of resources on servers of this group; $r_j(t)$ and Cap_j are the amount of resource j used at time t in the group considered, and the capacity of this resource, respectively.

5.4 Cost Formulation

In order to present a general cost for evaluating different proactive/reactive strategies we denote $y(k)$ the total energy demanded during a given time interval $[(k-1)T, kT)$, where $k = 1, 2, ...$; $u_p(k)$ and $u_r(k)$ the energy provided by servers at the beginning of this interval (proactive action), and at the end of this interval (reactive action), respectively. This general cost takes into account: (i) the energy cost paid to the DC energy provider; (ii) some penalty due to user dissatisfaction, because of a longer waiting time in case of underestimation; (iii) some penalty due to resource waste, because of an overestimation of the energy needed to serve the user requests; and (iv) at each step, k, the energy of reactive action fulfills $u_r(k+1) = e(k)$. Using these assumptions, we propose the following cost:

$$\boxed{C(k) = c_e y(k) + c_u max(0, e(k)) + c_o max(0, -e(k))} \qquad (5.5)$$

where c_e is the electricity price per Joule and the error $e = y - \hat{y}$. In this cost, we consider some penalty given by coefficient c_u representing the cost of 1 J of the energy that should be used to serve the waiting requests in the case of underestimation. It is a penalty due to user dissatisfaction because of a longer waiting time. We consider also some penalty c_o the cost of 1 J of

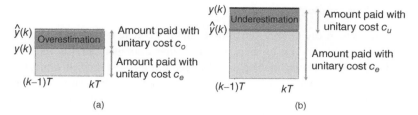

Figure 5.3 Costs paid for (a) over estimation and (b) underestimation.

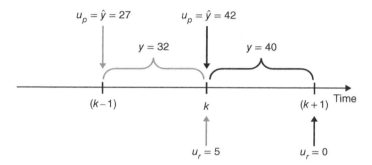

Figure 5.4 Example of energy predicted and consumed using proactive and reactive actions.

available but unused energy in the case of overestimation due to resource waste. The expression of this cost is given in (5.5).

The coefficients c_p, c_u, and c_o are greater than or equal to 0 and $c_e \neq 0$. Figure 5.3 depicts the cost paid in the case of overestimation (a) and underestimation (b).

5.4.1 Example

Figure 5.4 depicts an example of the energy consumed, the prediction, the prediction error, and the energy costs for step k in gray and step $k + 1$ in black. The DC management proceeds as follows: the amount of energy needed at step k is 32, whereas the prediction was $u_p = 27$. As an amount of energy of 27 was available at time $(k - 1)T$, an additional amount of $u_r = 5$ must be provided at time kT. The total cost at step k is $27c_e + 5c_u$. At step $k + 1$, the energy required to serve the new requests is 40. The additional amount of energy assigned proactively to the new period is available at time kT for step $k + 1$. It is given according to the prediction of $u_p = 42$. There is an overestimation of 2, hence $e = y - \hat{y} = y - u_p = -2$ and $u_r = max(0, e) = 0$. The total cost assigned to this period is then $40c_e + 2c_o$.

5.4.2 Methodology

Note that when the energy consumed remains constant from one step to the next, i.e. $y(k - 1) = y(k)$, there is no proactive management. In this case, this is equivalent to considering that the prediction is just the last value, i.e. $\hat{y}(k) = y(k - 1)$. Thus, in order to evaluate the energy saving with regard to the reactive management when using proactive actions, one must analyze the difference between the total cost achieved at step k when using intelligent predictors and the one obtained by using the last value as a prediction.

Since in most DCs (Reiss et al., 2012), servers of different generations and capacities coexist, our evaluation of the upper bound of energy cost saving takes into account server heterogeneity. The evaluation of energy cost savings is carried out according to the methodology, which comprises four steps:

- *Step 1*: Classify servers into groups such as all servers of the same group have the same features (e.g. memory, CPU, instruction execution time, energy consumption parameters).
- *Step 2*: Compute $y(k)$ the energy consumed by servers of a same group in each time interval $[(k - 1)T, kT)$ in the recent past. In this past time interval, we know the arrival time of each request served by servers in the group considered, the requested resources and, if finished, we also know its exact duration. Otherwise, we take the duration up to the end of time interval T. We then compute $y(k)$ the energy consumed by the servers of this group in the time interval $[(k - 1)T, kT)$ as explained in Section 5.3.2. We proceed similarly for each time interval in the recent past. We proceed similarly on each server group existing in the DC, taking into account server heterogeneity.
- *Step 3*: Compute the relative energy cost saving (RES) using proactive decisions, according to the definition given later.
- *Step 4*: Maximize RES using recent past data for predictions $\hat{y}$ for each group of servers. Considering *admissible predictors* as predictors that give a value of $RES > 0$, compute the upper bound UB for any admissible predictor. Intuitively, UB is the best RES that can be obtained on any data set by admissible predictors.

5.4.3 Relative Energy Cost Saving

Since $max(0, a) = \frac{a}{2} + \frac{|a|}{2}$ for any $a \in \mathcal{R}$, we get from the cost in (5.5)

$$C = \mathcal{E}\left[c_e y + \frac{c_u}{2}(e + |e|) + \frac{c_o}{2}(-e + |e|)\right]$$

$$= \mathcal{E}\left[c_e y + \frac{c_u - c_o}{2}e + \frac{c_u + c_o}{2}|e|\right] \tag{5.6}$$

where $\mathcal{E}[\cdot]$ means expectation. Let us define $\alpha = c_u/c_e$, $\beta = c_o/c_e$ and $J_{\alpha,\beta,T} = \frac{2C}{c_e}$. Since c_p, c_u, and c_o coefficients are greater than or equal to 0 and $c_e \neq 0$, then α and β are also greater than or equal to 0. We obtain:

$$\boxed{J_{\alpha,\beta,T} = 2\mathcal{E}[y] + (\alpha - \beta)\mathcal{E}[e] + (\alpha + \beta)\mathcal{E}[|e|]} \tag{5.7}$$

where $\mathcal{E}[y]$ and $\mathcal{E}[e]$ stand for the expected values of random variables y and error e, respectively.

In the case of perfect prediction, $\hat{y}(k) = y(k)$, we qualify as purely proactive the energy cost of this group of servers. Then, taking into account that (i) the cost, $J_{\alpha,\beta,T}$, can be reduced by using good predictions and (ii) the cost, $L_{\alpha,\beta,T}$, due to the last value prediction, $\hat{y}(k) = y(k-1)$ is a purely reactive cost, we are interested in evaluating the relative saving that can be achieved when using proactive action. Let us define the relative energy cost Saving $RES_{\alpha,\beta,T}$ using proactive action in the group considered as:

$$RES_{\alpha,\beta,T} = \frac{L_{\alpha,\beta,T} - J_{\alpha,\beta,T}}{L_{\alpha,\beta,T}} = 1 - \frac{J_{\alpha,\beta,T}}{L_{\alpha,\beta,T}} \tag{5.8}$$

The greater $RES_{\alpha,\beta,T}$ is, the higher the cost reduction is. By replacing the expressions of $L_{\alpha,\beta,T}$ and $J_{\alpha,\beta,T}$, we get:

$$\boxed{RES_{\alpha,\beta,T} = 1 - \frac{2\mathcal{E}[y] + (\alpha - \beta)\mathcal{E}[e] + (\alpha + \beta)\mathcal{E}[|e|]}{2\mathcal{E}[y] + (\alpha - \beta)\mathcal{E}[d] + (\alpha + \beta)\mathcal{E}[|d|]}} \tag{5.9}$$

where $d(k) = y(k) - y(k-1)$.

The optimal predictor for the group considered is that which maximizes the relative energy cost saving $RES_{\alpha,\beta,T}$, or equivalently minimizes $J_{\alpha,\beta,T}$ to achieve the best energy cost savings.

$$Optimal\ predictor = \arg\min_{\hat{y}} \left\{J_{\alpha,\beta,T}\right\}. \tag{5.10}$$

Remarks. For the detail of the predictor design, see Milocco et al. (2020). The predictor that minimizes the cost $J_{\alpha,\beta,T}$ is different from the one that minimizes the mean square error (MSE) as shown hereafter:

$$J_{\alpha,\beta,T} = 2\mathcal{E}[y] + (\alpha - \beta)\mathcal{E}[e] + (\alpha + \beta)\mathcal{E}[|e|]$$
$$= 2\mathcal{E}(y) + (\alpha + \beta)\left(\frac{\alpha - \beta}{\alpha + \beta}\mathcal{E}(e) + \mathcal{E}|e|\right) \tag{5.11}$$

Then, we get in general:

$$\arg\min_{\hat{y}} \left\{ J_{\alpha,\beta,T} \right\} = \arg\min_{\hat{y}} \left\{ \mathcal{E}|e| + \frac{\alpha - \beta}{\alpha + \beta} \mathcal{E}[e] \right\} \tag{5.12}$$

$$\neq \arg\min_{\hat{y}} \left\{ \mathcal{E}(e^2) \right\} = MSE$$

However, when $e = 0$ both coincide.

5.4.4 Upper Bound Computation

We want to establish the upper bound on the relative energy cost savings that can be obtained, whatever the predictor considered. Notice, however, that this bound depends on the data set considered. Suppose we wish to predict the value of the sample $y(k)$ using a linear combination of infinite past samples $y(k-1), y(k-2), \dots$. The optimal one-step-ahead prediction error (i.e. minimizing $\mathcal{E}[e^2]$) is an uncorrelated sequence with constant power spectrum density (PSD). The minimum mean-square error (MMSE) is given by the Szegö–Kolmogorov theorem which states that

$$MMSE = \mathcal{E}[e^2] = exp\left(\int_0^{2\pi} \ln(\Phi(j\omega)d\omega \right) \tag{5.13}$$

where j is the complex number $j^2 = -1$, ω is the frequency in radians per second, and $\Phi(j\omega)$ is the PSD of $y(k)$ Brockwell and Davis (1986).

The simplest upper bound is obtained when the prediction error is null. However, this property is never met by a realistic predictor. That is why we propose a realistic upper bound that can be reached by a predictor. To find this upper bound, we need to consider the minimum of $(\alpha - \beta)\mathcal{E}[e] + (\alpha + \beta)\mathcal{E}[|e|]$ in Eq. (5.7). Since $\mathcal{E}[e] \leq \mathcal{E}[|e|]$, the minimum is obtained when $e(k)$ is the negative constant value for all k. In this case $\mathcal{E}[e] = -\mathcal{E}[|e|] = -\mathcal{E}[e^2]^{1/2} = -\gamma$, where γ is MMSE given by Eq. (5.13), we get that the minimum is $(\alpha - \beta)(-\gamma) + (\alpha + \beta)\gamma$ in Eq. (5.7), leading to the following value for the upper bound $UB_{\alpha,\beta,T}$ of $RES_{\alpha,\beta,T}$:

$$UB_{\alpha,\beta,T} = 1 - \frac{2(\mathcal{E}[y] + \beta\gamma)}{2\mathcal{E}[y] + (\alpha - \beta)\mathcal{E}[d] + (\alpha + \beta)\mathcal{E}[|d|]} \tag{5.14}$$

5.5 Application to a Real DC

In this section, a numerical comparison is carried out using experimental data collected from a real DC that allow us to obtain the maximum levels of cost reduction. Thus, the performance of different predictors can be quantified for the DC considered.

5.5.1 Generalities

5.5.1.1 Selection of the Sampling Interval

The value of T is a trade-off: small values of T make the DC management more adaptive to workload changes, whereas large values of T decrease the management overhead. However, switching a server off and on has a cost. Hence, the minimum value of T, denoted T_{min} can be obtained as in Dabbagh et al. (2015) by considering the smallest time interval for which switching off a server and then switching it on in the next time interval has a null energy cost. Too large a value of T is not desirable either, as the delay to serve the requests may become far too large. If the predicted value is underestimated, some users might have to wait for too long until the next configuration time. If the predicted value is overestimated, extra energy is consumed for too long. Hence, the maximum value of T, denoted T_{max}, is determined by the maximum waiting time acceptable by users in the worst case.

5.5.1.2 Selection of Possible Values for the Costs

Using the cost formulation (5.6), the value of the coefficients c_e, c_u, and c_o can be obtained from historical records, since the true and estimated energies, and also the total energy cost in each time interval T are known. For N consecutive time intervals in the recent past, the following formulation holds:

$$\begin{bmatrix} C(1) \\ \vdots \\ C(N) \end{bmatrix} = c_e \begin{bmatrix} y(1) \\ \vdots \\ y(N) \end{bmatrix} + c_u \begin{bmatrix} \frac{e(1)+|e(1)|}{2} \\ \vdots \\ \frac{e(N)+|e(N)|}{2} \end{bmatrix} - c_o \begin{bmatrix} \frac{e(1)-|e(1)|}{2} \\ \vdots \\ \frac{e(N)-|e(N)|}{2} \end{bmatrix} \quad (5.15)$$

where $C(k)$ denotes the total energy cost for time interval $[(k-1)T, kT)$. The coefficients c_e, c_u, and c_o are computed using the least squares method.

5.5.1.3 Dynamic Capacity Provisioning Based on Energy Prediction

The basic principle of energy saving in a DC lies in turning off servers which are not being used during a time greater than or equal to T_{min}. More generally, how to provide DCP (Zhang et al., 2014) when only the energy that should be consumed in the next time interval is predicted.

We distinguish two cases. If in the previous interval, the predicted energy $\hat{y}(k-1)$ is greater than or equal to the energy actually consumed, $y(k-1)$, the energy that is provided at time $(k-1)T$ for the next time interval $[(k-1)T, kT)$ is equal to the prediction $\hat{y}(k)$.

On the other hand, if in the previous interval, the predicted energy $\hat{y}(k-1)$ is strictly less than the energy actually consumed, $y(k-1)$, some requests are still waiting to be served at time $(k-1)T$. Consequently, the energy that is provided at time $(k-1)T$ for the next time interval is equal to the sum of the prediction $\hat{y}(k)$ and the amount of energy needed to compute the waiting requests, $e(k-1)$.

To summarize both cases, the energy that is provided at time $(k-1)T$ for the next time interval is equal to $\hat{y}(k) + max(0, e(k-1))$.

In the single resource case, we define $m(k)$, the minimum number of servers on in the group considered for the time interval $[(k-1)T, kT)$ to provide the energy $\hat{y}(k) + max(0, e(k-1))$ as:

$$m(k) = \left\lceil \frac{\hat{y}(k) + max(0, e(k-1))}{P_{peak}TC_{ap}} \right\rceil \tag{5.16}$$

where P_{peak} is the power consumed by a server of this group at 100% load. Its value is given by:

$$P_{peak} = \frac{P_{idle}\mu + \eta}{C_{ap}} \tag{5.17}$$

In the case of multiple resources, we predict the energy consumed by each resource present in the group considered. More precisely, we compute the utilization factor of this resource using the second term of the sum in Eq. (5.1) This utilization factor takes into account the amount of resource required to serve the waiting requests at the end of the current time interval, if any. We then deduce which resource is the bottleneck resource. Finally, we compute the number of machines needed to provide this energy for the next time interval.

5.5.2 Application to a Google Dataset

The results reported in this section have been obtained using the Google data set described in Reiss et al. (2011) and collected in an operational DC over a period of 29 days. According to our methodology, we first classify servers into groups, where servers of the same group have the same features. This classification is given by Table 5.2, where the CPU capacity is normalized with regard to the most powerful CPU, whereas the memory capacity is normalized with regard to the largest memory. Table 5.2 shows that 92.65% of servers (i.e. all servers belonging to the first seven groups) have the same CPU capacity of 0.5. That is why in the following, we consider a single group with a CPU capacity of 0.5.

5.5.2.1 Energy Computation

The energy parameters used for the Google DC are those given in Dabbagh et al. (2015), where the power consumed by a server increases linearly from

Table 5.2 The server groups in the Google DC

Group	CPU capacity	Memory capacity	Percentage of servers (%)
1	0.5	0.5	53.45
2	0.5	0.25	30.73
3	0.5	0.75	7.97
4	0.5	0.125	0.43
5	0.5	0.03	0.039
6	0.5	1	0.031
7	0.5	0.06	0.007
8	1	1	6.31
9	1	0.5	0.023
10	0.25	0.25	0.98

$P_{idle} = 300$ W at a null workload to $P_{peak} = 600$ W at a 100% workload. We define the energy consumed by any server of the group considered at its peak load during T as,

$$E_{peak,T} = P_{peak} TCap \tag{5.18}$$

leading to,

$$y(k) = P_{peak} \int_{(k-1)T}^{kT} r(t)dt \tag{5.19}$$

We apply the proposed methodology to evaluate the improvements of using proactive management compared to reactive management in the range of $T_{min} = 60$ seconds to $T_{max} = 3600$ seconds (i.e. 1 hour). This methodology allows us to know what the benefits would be of using the proposed predictors as well as the upper bound of these benefits. We consider two types of predictors: linear and nonlinear. We choose the ARMA model as a linear predictor. The recursive algorithm used by the linear predictor works on a sliding window of past samples. We also select a nonlinear predictor based on the conditional probability. Figure 5.5 depicts the relative energy consumed by the DC, its prediction by the linear optimal predictor and the nonlinear pptimal predictor for $T = 3600$ seconds and $T = 60$ seconds. From Figure 5.5, it is possible to see that for $T = 3600$ seconds the energy is smoother than for $T = 60$ seconds, and that both predictors tend to overestimate the energy because the underestimation cost c_u is higher than overestimation one c_o.

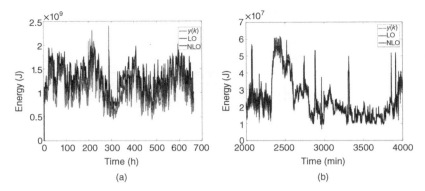

Figure 5.5 True relative energy consumed vs. its linear and nonlinear predictions for $T = 3600$ seconds (a) and $T = 60$ seconds (b) for $c_e = c_o = 1$ and $c_u = 9$.

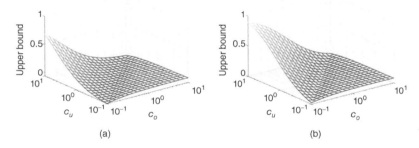

Figure 5.6 Upper bound for $T = 60$ seconds (a) $T = 3600$ seconds (b).

5.5.2.2 Evaluation of the Upper Bound

The evaluation of the upper bound is performed according to Eq. (5.14), for different values of c_e, c_o, and c_u, and for time intervals $T = 60$ and $T = 3600$ seconds, using the energy needed to serve the CPU requests. The results are depicted in Figure 5.6. The possible savings are between 1% and 85% depending on the values of c_o/c_u, and T. It can be seen that the highest values are found for decreasing values of c_o/c_u and increasing values of T. This is due to the fact that when the underestimation cost increases, the predictor tends to overestimate; hence, the upper bound increases considerably. For $c_o = c_u$, we can observe in Figure 5.6 that the upper bound is near to zero. This is because the prediction error $e(k)$ is similar to the difference $d(k)$ in Eq. (5.14); therefore, the relative energy cost saving $RES_{\alpha,\beta,T}$ is low for the Google data set studied.

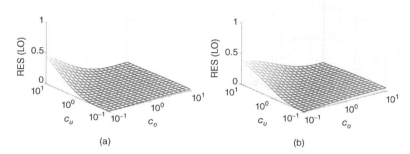

Figure 5.7 RES for CPU requests using the linear optimal predictor for $T = 60$ seconds (a) and $T = 3600$ seconds (b).

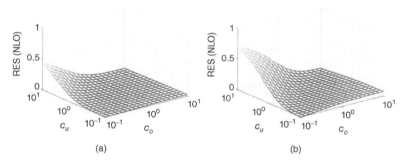

Figure 5.8 RES for CPU requests using the nonlinear optimal predictor for $T = 60$ seconds (a) and $T = 3600$ seconds (b).

5.5.2.3 Computation of the Relative Energy Cost Saving

The evaluation of $RES_{\alpha,\beta,T}$ is performed for different sets of admissible α, β, and T using the energy needed to serve the CPU requests, leading to the results depicted in Figure 5.7 for the linear optimal predictor and Figure 5.8 for the nonlinear optimal predictors. With both predictors, energy cost savings are obtained only for values of $c_o/c_u \leq 1$. Notice also that the nonlinear optimal predictor gets much closer to the upper bound than the linear optimal predictor.

Small values of T give lesser values of $RES_{\alpha,\beta,T}$, but the relative energy cost savings strongly depend on the value of c_o/c_u. For decreasing values of $c_o/c_u < 1$, $RES_{\alpha,\beta,T}$ increases with all the predictors. Although $RES_{\alpha,\beta,T}$ increases with larger values of T, we consider no value larger than 3600 seconds to minimize user dissatisfaction caused by a long waiting time before job execution in the case of underestimation.

5.5.2.4 Discussion of Results

This example shows that by applying the methodology proposed, it is possible to improve costs achieved by Google by up to 85%, strongly depending on the value of c_o/c_u. This example highlights the importance of knowing the upper bound, because it helps the DC manager to compare the predictor used with regard to the upper bound.

The results showed that the value of c_o/c_u has a strong impact on the relative energy cost savings one can get. More generally, there exists a trade-off between energy savings and QoS. If one tries to improve the QoS, more energy is spent because more servers are on, and vice versa. However, in our approach, energy fulfills a perfect balance at each step, because the proactive and reactive action strategy chosen keep perfect balance of energy at each step. Note that from Eq. (5.9) when $c_o << c_u$, the worst case of $RES(e)$ for a constant error $e = \pm\gamma$ fulfills $RES(-\gamma) > RES(\gamma)$. Therefore, RES can be seen as a measure of the QoS. For the DC taken as an example, we see that when T increases, RES is higher due to the proactive action. The best value of T should also be chosen to provide the best value of $RES_{\alpha,\beta,T}$.

To summarize, the value chosen for T is obtained by successive refinements. First, it should belong to the interval $[T_{min}, T_{max}]$. Second, as previously pointed out, it is a trade-off between energy saving and QoS. Third, for a given value of β, the value of T chosen by the DC manager maximizes the value of $RES_{\beta,T}$ for the predictor considered. Hence, according to the results of this study (Figures 5.7 and 5.8), this value optimizes the energy saving for the QoS required.

To mitigate the effect of underestimation, Dabbagh et al. (2015) propose two solutions. The first one tries to avoid underestimation, by adding a safety margin to the prediction of the number of servers. This solution may lead to overestimation which is not energy efficient. To limit overestimation, the safety margin is dynamically tuned according to the prediction error. The goal of the second solution is to limit the DC latency to recover from this underestimation. To react more promptly (e.g. every 10 seconds for a time interval $T = 60$ seconds), every $T/10$ for instance, an underestimation check is done. If some underestimation is detected, some servers are turned on to reduce user waiting time. In our case, the cost minimization automatically chooses the optimal value to avoid underestimation.

5.6 Conclusion

The purpose of this chapter is to evaluate the energy cost savings that can be achieved by a proactive DC management. Such a management consists in periodically configuring the DC by turning some servers on or off to provide an amount of energy according to the energy requested and the prediction error, using the proposed cost. The criteria used for the cost are (i) the energy cost paid to the DC energy provider, (ii) some penalty due to user dissatisfaction, because of a longer waiting time in case of underestimation, and (iii) some penalty due to resource waste, because of an overestimation of the energy needed to serve the user requests. Two predictors are proposed: a linear one based on the ARMA model and a nonlinear one using the conditional probability density function. These predictors maximize the relative energy cost saving.

We determined the tight upper bound of the relative energy cost savings valid whatever the predictor used. Knowing the upper bound makes it possible to decide on the suitability or not of using proactive action. Therefore, this paper proposes a methodology that helps to choose the best strategy (i.e. that maximizes the savings) between possible proactive strategies.

The methodology proposed is generic and can be applied to any DC that meets the assumptions made in this paper. These assumptions are realistic and can be met by many DCs. As an example, we have used the data set collected over 29 days in an operational DC of Google. By applying this methodology on this data set, an improvement up to 85% can be obtained, leaving room for multiple optimizations.

References

M. Alam, K.A. Shakil, and S. Sethi. Analysis and clustering of workload in Google cluster trace based on resource usage. *arXiv*, 2015.

O. Beaumont, L. Eyraud-Dubois, and J.A. Lorenzo-Del-Castillo. Analyzing real cluster data for formulating allocation algorithms in cloud platforms. *International Symposium on Computer Architecture and High Performance Computing*, 2014.

P.J. Brockwell and R.A. Davis. *Time Series: Theory and Methods.* Springer-Verlag, 1986.

S. Caires and J.A. Ferreira. On the non-parametric prediction of conditionally stationary sequences. *Statistical Inference for Stochastic Processes*, 2005. 8. 151–184.

R. Cao, Z. Yu, T. Marbach, J. Li, G. Wang, and X. Liu. Load prediction for data centers based on database service. *Computer Software and Applications Conference*, 2018.

F. Chen, J. Grundy, Y. Yang, J.-G. Schneider, and Q. He. Experimental analysis of task-based energy consumption in cloud computing systems. *ACM/SPEC ICPE*, 2013.

M. Dabbagh, B. Hamdaoui, M. Guizani, and A. Rayes. Energy-efficient resource allocation and provisioning framework for cloud data centers. *IEEE Transactions on Network and Service Management*, 2015.. 12. 377–391.

M. Dayarathna, Y. Wen, and R. Fan. Data center energy consumption modeling: a survey. *IEEE Communications Surveys & Tutorials*, 2016. 18. 732–794.

C. Delimitrou and C. Kozyrakis. Quasar: resource-efficient and QoS-aware cluster management. *International Conference on Architectural Support for Programming Languages and Operating Systems*, 2014.

S. Di, D. Kondo, and W. Cirne. Host load prediction in a Google compute cloud with a Bayesian model. *International Conference on High Performance Computing, Networking, Storage and Analysis*, 2012.

T. Hastie, R. Tibshirani, and J. Friedman. *The Elements of Statistical Learning: Data Mining, Inference, and Prediction*. Springer, 2009.

B. Liu, Yinan Lin, and Y. Chen. Quantitative workload analysis and prediction using Google cluster traces. *IEEE Conference on Computer Communications Workshops*, 2016.

R. Milocco, P. Minet, E. Renault, and S. Boumerdassi. Evaluating the upper bound of energy cost saving by proactive data center management. *Inria Research Report*, 2020.

P. Minet, E. Renault, I. Khoufi, and S. Boumerdassi. Analyzing traces from a Google data center. *International Wireless Communications and Mobile Computing Conference*, 2018.

E. Pacini, L. Iacono, C. Mateos, and C.G. Garino. *A Bio-Inspired Datacenter Selection Scheduler for Federated Clouds and Its Application to Frost Prediction*. Springer-Verlag, 2018.

J. Prevost, K. Nagothu, B. Kelley, and M. Jamshidi. Prediction of cloud data center networks loads using stochastic and neural models. *International Conference on System of Systems Engineering*, 2011.

C. Reiss, J. Wilkes, and J.L. Hellerstein. Google cluster-usage traces: format + schema. Google Inc. Technical Report, 2011.

C. Reiss, A. Tumanov, G.R. Ganger, R.H. Katz, and M.A. Kozuch. Heterogeneity and dynamicity of clouds at scale: Google trace analysis. *ACM Symposium on Cloud Computing (SoCC)*, 2012.

A. Wolke, M. Bichler, and T. Setzer. Planning vs. dynamic control: resource allocation in corporate clouds. *IEEE Transactions on Cloud Computing*, 2014. 4. 322–335.

P. Xiao, Z. Hu, D. Liu, G. Yan, and X. Qu. Virtual machine power measuring technique with bounded error in cloud environments. *Journal of Network and Computer Applications*, 2013.. 36. 818–828

M.S. Yoon, A.E. Kamal, and Z. Zhu. Adaptive data center activation with user request prediction. *Computer Networks*, 2017. 122. 191–204

S. Yousif and A Al-Dulaimy. Clustering cloud workload traces to improve the performance of cloud data centers. *World Congress on Engineering*, 2017.

Q. Zhang, M.F. Zhani, R. Boutaba, and J.L. Hellerstein. Dynamic Heterogeneity-aware resource provisioning in the cloud. *IEEE Transactions on Cloud Computing*, 2014. 2. 14–28.

Index

Management of Data Center Networks, First Edition. Edited by Nadjib Aitsaadi.
© 2021 The Institute of Electrical and Electronics Engineers, Inc.
Published 2021 by John Wiley & Sons, Inc.

IEEE Press Series on
Networks and Services Management

The goal of this series is to publish high quality technical reference books and textbooks on network and services management for communications and information technology professional societies, private sector and government organizations as well as research centers and universities around the world. This Series focuses on Fault, Configuration, Accounting, Performance, and Security (FCAPS) management in areas including, but not limited to, telecommunications network and services, technologies and implementations, IP networks and services, and wireless networks and services.

Series Editors:
Dr. Veli Sahin
Dr. Mehmet Ulema

---

1. *Telecommunications Network Management into the 21st Century*
   Edited by Thomas Plevyak and Salah Aidarous
2. *Telecommunications Network Management: Technologies and Implementations: Techniques, Standards, Technologies, and Applications*
   Edited by Salah Aidarous and Thomas Plevyak
3. *Fundamentals of Telecommunications Network Management*
   Lakshmi G. Raman
4. *Security for Telecommunications Network Management*
   Moshe Rozenblit
5. *Integrated Telecommunications Management Solutions*
   Graham Chen and Qinzheng Kong
6. *Managing IP Networks: Challenges and Opportunities*
   Edited by Thomas Plevyak and Salah Aidarous
7. *Next-Generations Telecommunications Networks, Services, and Management*
   Edited by Thomas Plevyak and Veli Sahin
8. *Introduction to IP Address Management*
   Timothy Rooney
9. *IP Address Management: Principles and Practices*
   Timothy Rooney

Printed and bound by CPI Group (UK) Ltd, Croydon, CR0 4YY